BRITISH COLUMBIA
WINE
COUNTRY

BRITISH COLUMBIA
WINE
COUNTRY

text **John Schreiner** photography **Kevin Miller**

whitecap

Text copyright © 2003 by John Schreiner
Photographs copyright © 2003 by Kevin Miller/kevinmiller-photography.com
Whitecap Books

Third printing, 2005

All rights reserved. No part of this publication may be reproduced, stored in a retrieval system, or transmitted in any form or by any means, electronic, mechanical, photocopying, recording or otherwise, without prior written permission from the publisher. For more information, contact Whitecap Books Ltd., 351 Lynn Avenue, North Vancouver, British Columbia, Canada, V7J 2C4.

Edited by Elaine Jones
Proofread by Marial Shea
Cover design by Jane Lightle
Interior design by Margaret Lee /bamboosilk.com
Maps by Jacqui Thomas

Printed and bound in Canada

National Library of Canada Cataloguing in Publication Data
Schreiner, John, 1936–
    British Columbia wine country / John Schreiner; photographer, Kevin Miller.

    Includes bibliographical references and index.
    ISBN 1-55285-336-5

    1. Wine and wine making—British Columbia. 2. Wineries—British Columbia. 3. Viticulture—British Columbia. I. Title.
TP559.C3S335 2002   641.2'2'09711   C2002-911390-3

The publisher acknowledges the support of the Canada Council and the Cultural Services Branch of the Government of British Columbia in making this publication possible. We acknowledge the financial support of the Government of Canada through the Book Publishing Industry Development Program for our publishing activities.

PREVIOUS PAGE:
A BOUNTIFUL HARVEST PREPARED FOR THE JOURNEY TO THE WINERY.

PAGES 6-7: **ONE OF THE MOST BREATHTAKING PROPERTIES IN WINE COUNTRY IS MISSION HILL'S NARAMATA VINEYARD ESTATE, FORMERLY PARADISE RANCH, WHICH PERCHES DRAMATICALLY ABOVE OKANAGAN LAKE.**

THIS BOOK IS DEDICATED TO THE MEMORY OF VICTOR MANOLA AND FRANK SUPERNAK, WHO BOTH PERISHED IN AN ACCIDENT IN NOVEMBER 2002 WHILE FINISHING THE VINTAGE AT THE RECENTLY COMPLETED SILVER SAGE WINERY. MANOLA WAS THE CO-FOUNDER OF SILVER SAGE. SUPERNAK, A RESPECTED OKANAGAN WINEMAKER SINCE 1984, RECENTLY MADE WINE AT HESTER CREEK AND BLASTED CHURCH AND WAS CONSULTING AT CHALET ESTATES AND SILVER SAGE.

ACKNOWLEDGMENTS

A book like this would not be possible without the generosity of winery owners and winemakers of British Columbia who, once again, interrupted busy lives to answer endless questions about everything from their places of birth to their philosophies of wine. A special thanks goes to the many — they know who they are — who welcomed me informally at lunch or dinner with their families. I also would like to recognize the support and forbearance of Marlene, my wife, who, strangely enough, has come to take most of her vacations in wine country.

—JOHN SCHREINER

My wife Sunantha for all of her support throughout this project, especially the beginning, my late grandfather Andrew Orason who preferred a cold "Bud," my mother Vicki Tyndall for putting up with me all of these years, same goes for my father Roger Miller. Dave and Ladda Green for giving me a home away from home, Mabel (Dorothy) Phoon for steering me straight, Andrew Waddy the quail hunter, Eric Kerwin for keeping his dogs on a leash, Kenneth Hadzima for graciously handing over the remote so that I could watch the weather channel, Ron Sangha the creator, HRH Richard Peitz for being himself, Billy and Megan Goddard for the deadhead that wasn't, Anji Smith for all of the black & white prints, Tim Pawsey for all of his patient hours on the phone, Christine Colleta for her enthusiasm, Robert McCullough at Whitecap Books for being a wine enthusiast, and Michael Buckley for his invaluable insight. I am also very thankful to Simon N. Cumming, Megan Moyle, Eric von Krosigk, Wayne and Joanne Finn, Blair Baldwin, Leanne McDonald, Dave Gamble, Corrine Kovalsky, Dawn Antle, Fritz Hollenbach, Lawrence Page, Walter Gehringer, Martha Hivon, Jim and Devie Cairney, Pieter Koster and Dave Laamanen at Canadian Helicopters, and the many people in the wine industry who shared their knowledge, support and time making this project possible.

—KEVIN MILLER

CONTENTS

11   *TO BEGIN WITH...*

16   VANCOUVER ISLAND VINEYARDS

36   GULF ISLANDS VINEYARDS

48   FRASER VALLEY VINEYARDS

60   WINEMAKERS OFF THE BEATEN PATH

72   OLD VINES COUNTRY — THE VINEYARDS OF KELOWNA

88   THE VINEYARDS OF MOUNT BOUCHERIE

100 FROM SUMMERLAND TO PEACHLAND

110 PENTICTON AND THE NARAMATA BENCH

130 VINEYARDS OF OKANAGAN FALLS

142 THE GOLDEN MILE

160 THE BLACK SAGE ROAD NEIGHBOURHOOD

174 OSOYOOS LAKE BENCH

180 THE SIMILKAMEEN VALLEY

188 *INDEX*

192 *ABOUT THE AUTHOR AND PHOTOGRAPHER*

TO BEGIN WITH...

"WINE IS NOT A PROFESSION, IT IS AN OBSESSION."
—GEORGE HEISS SR.
GRAY MONK

In October 1981, four months after buying Mission Hill Vineyards (as it was then called), Anthony von Mandl hosted a winery luncheon for about 100 people, most of them from the Okanagan, and gave his remarkable *I have a dream* speech. It is doubtful that his audience took his grand vision very seriously, considering that British Columbia wines then were losing substantial ground to imported wines. Even von Mandl did not keep a copy of his speech. However, it remained in my own files. Several years ago, when it was obvious that the dream had been achieved, I gave von Mandl photocopies of his text.

In 1981, there were 14 wineries in British Columbia. Many were struggling to retain consumers with off-dry wines from hybrid grapes whose labels — like Calona's Schloss Laderheim — mimicked the European labels they were trying to hold off. The new estate wineries, notably Gray Monk and Sumac Ridge, were creating excitement with their wines. The rest of the industry was not well-respected. Against that background, the still boyish von Mandl presumed to tell his guests that "this majestic valley is resting on the threshold of being an economic giant." He might just as easily have told them that Ogopogo exists.

"When I look out over the valley, I see world-class vinifera vineyards winding their way down the valley; numerous estate wineries, each distinc-

OPPOSITE: **COOL CELLAR OF A MODERN WINERY WITH GLEAMING STEEL TANKS AND FRAGRANT OAK BARRELS.**

tively different; charming inns and bed and breakfast cottages seducing tourists from around the world while intimate cafés and restaurants captivate the visitor in a magical setting," von Mandl said. "It is an undisputed fact that the interest in wine over the past ten years has been explosive. The Napa Valley in California has taken advantage of this in a remarkable way, creating a billion-dollar industry. We in the Okanagan have not truly even begun. The first step is to set up wine and vineyard tours of the Okanagan."

At the time, von Mandl's guests would have regarded the comments as optimistic at best. Remarkably, his vision has been more than fully realized. By the summer of 2002, British Columbia had 71 licensed wineries, five licensed fruit wineries and 16 wineries with pending licenses. Another dozen or so were not far enough along to reach the license-pending status but definitely were under development at that time. Altogether, there are about 100 producers and wineries in waiting. Visitors now throng to the wine regions, finding vineyard tours, restaurants at wineries, and all the other amenities that von Mandl foresaw in 1981. His own winery, tired and full of fruit flies when he bought it, is a spectacularly rebuilt cathedral to wine, with product to match. The wines of Mission Hill and its peers are internationally competitive in quality and are listed with pride in virtually every restaurant and café in British Columbia.

Most of this has happened since 1988. The free trade agreement became effective the following year. Almost overnight, British Columbia wineries lost preferential status in liquor stores. Until then, imported wines, if they were allowed in at all, were always priced at a premium to British Columbia

OPPOSITE: **PENTICTON'S WINE INFORMATION CENTRE DISPENSES WINE TOURING HELP AND SELLS SOME OF THE OKANAGAN'S BEST WINES.**

THE WINE COUNTRY'S LONG, WARM SUMMER IS IDEAL FOR SAVOURING FOOD AND WINES IN THE VINEYARDS.

wines. The free trade agreement required an open market with practically no price discrimination. Suddenly, domestic wine made from mediocre grapes such as De Chaunac and Okanagan Riesling had no future. Faced with free trade, two-thirds of the 3,000 acres (1,214 hectares) of vineyards in the Okanagan were pulled out after the 1988 harvest.

WINERY OWNER JIM WYSE AND A VISITOR SAMPLE BURROWING OWL WINES FROM THE BARREL.

The remaining grapes were primarily quality vinifera. It was on these vineyards that the industry relaunched itself.

In 1991, the Vintners Quality Alliance program was introduced. Consumers interpreted VQA as an assurance of quality because the wines, all made from British Columbian–grown grapes, are tasted and approved by the VQA panel before being sold. At the same time as VQA wines began appearing, a new group of small wineries — then called farm gate wineries — began opening. Both VQA and the farm wineries created fresh excitement about British Columbia wines. The Okanagan Wine Festival, launched as a fall tourist promotion in 1982, now began to draw serious numbers of visitors to the Okanagan for the vintage. By 1992, it was clear that winemaking would survive in British Columbia. The uprooted vineyards were acquired by new owners who imported quality varietal vines from France. By 2001, twice as many acres of vineyard were planted or were about to be planted as even existed in 1981. It is a cliché, but still true, that good wine is made primarily in the vineyard.

Jeff Martin, one of the owners at La Frenz winery near Penticton, remonstrates that wine is also made by winemakers. That comment identifies the other major change in British Columbia wineries since von Mandl's speech. There were few winemakers or vineyard managers then with international training and experience and there was little winemaker training being done in Canada. Today, there is a healthy level of expertise at work in both the vineyards and the wineries, with winemakers trained in leading wine schools of Europe, Australia, New Zealand, South Africa and the United States. As well, Okanagan University College has been turning out capable winery technicians since 1997.

Some wineries in British Columbia are still struggling; that is the nature of any business. But this is a far cry from 1981 when *most* were struggling. British Columbia wines now command respect at home and abroad. It has been a remarkable transition in a short time.

This book sets out to celebrate that achievement. For the purposes of the book, British Columbia wine country has been divided into 13 regions. This is not intended to impose an appellation system on British Columbia, since it will take another generation or two to define appellations, where appropriate. With the exception of chapter four, each chapter deals with a different geographic cluster of wineries, conveniently arranged for wine touring as much as for the similarity of the local terroir.

WINE SPEAK

Acidity: This natural tartness in grapes and other fruit contributes to vibrant flavours.

Appellation: The geographical definition of a wine region. Currently, British Columbia's only appellations are Okanagan Valley, Similkameen Valley, Fraser Valley and Vancouver Island.

Botrytis: This is a fungus that attacks grape skins. In favourable conditions (misty mornings, dry afternoons), it dehydrates grapes, allowing the production of intense dessert wines. In unfavourable conditions (too much rain), it rots the grapes.

Brix: A measure of sugar in grapes: one degree Brix equals 18 grams of sugar per litre. Mature grapes typically are 21 to 25 Brix, equating to 11 to 13 percent alcohol after fermentation.

Cidery: An establishment that makes apple cider.

Clone: The mutation of a species. Growers select and propagate clones selected for such desirable qualities as early ripening, vivid flavour, deep colour. Several clones of the same variety often grow in the same vineyard to produce more complex wines when blended.

Commercial winery: A British Columbia winery with a commercial license is permitted to bottle and sell wine made with imported wine or fruit.

Estate winery: A winery with vineyards of its own. The 1978 regulation, now obsolete, required estate wineries to have at least 20 acres (eight hectares) of vines. Under current regulations, these wineries are licensed as land-based wineries. They only use British Columbia grapes or fruit.

Farm gate winery: This license was created in 1989 for wineries with vineyards too small to qualify as estate wineries. These are now included with the land-based wineries.

Fermentation: The natural process in which yeast converts sugar to alcohol.

Fruit wine: Wines made from fruit other than grapes.

Hybrid: Grape varieties developed, typically, by crossing European varieties with native North American varieties. The plant breeder's objectives include developing varieties that ripen early or resist disease or are winter-hardy.

Labrusca: A family of grapes — *Vitis labrusca* — native to North America and not suitable for table wines. Best known labrusca variety is Concord.

Meadery: An establishment that makes mead, a fermented alcoholic beverage made with honey.

Meritage: This word, which rhymes with heritage, was created in California to identify blends made with Bordeaux grape varieties. White Meritage is a blend of Sauvignon Blanc and Sémillon; red Meritage typically is a blend of Merlot, Cabernet Sauvignon, Cabernet Franc and sometimes Malbec or Petite Verdot. Wineries using the Meritage label subscribe to quality standards of the Meritage Association.

Must: Unfermented grape pulp or juice.

Organic: A technique for growing grapes (and other plants) without using chemicals such as pesticides, herbicides or commercial fertilizers.

Tannin: A compound in the skin and seeds of grapes that is essential in providing substance and backbone to red wines. When there is excessive tannin, wines are hard on the palate and have an astringent, bitter aftertaste, not unlike overly strong tea.

Terroir: Borrowed from the French, this term encompasses the entire environment — soil, climate, aspect — that makes a vineyard special.

Varietal: When the grape variety in the wine is named on the label (such as Chardonnay, Merlot), the wine is a varietal.

Vinifera: This is the species grape — *Vitis vinifera* — that has produced the classic wine grapes that have spread from the vineyards of Europe to wherever fine wine is made.

VQA or Vintners Quality Alliance: All British Columbia wines bearing the VQA seal must be made from British Columbia grapes. The wines are all screened by a professional tasting panel. Wines found faulty cannot be sold as VQA wines. About half of British Columbia wineries — including very competent producers — have chosen, for various reasons, not to submit wines for the VQA seal.

VANCOUVER ISLAND VINEYARDS

"THERE ARE ONLY TWO TYPES OF WINE:
THE WINE YOU LIKE AND THE WINE YOU DON'T LIKE."

—JOE BUSNARDO
DIVINO ESTATE WINERY

To make wine on Vancouver Island, it helps to be seduced by wine growing because the challenge is so considerable. "We are in a marginal grape-growing area," admits Roger Dosman, the proprietor of Alderlea Vineyards just northeast of Duncan. He is here by choice, having spent four years researching potential vineyards after selling a successful autobody repair business in Vancouver in 1988. His choice is hardly unique. Scottish-born John Kelly, now one of the owners of Glenterra Vineyards at Cobble Hill, looked at properties as distant as France when the winemaking passion came upon him. Joe Busnardo, the first successful grower of European grapes in the Okanagan, relocated his Divino Estate Winery in 1996 from there to a new vineyard overlooking the island highway. Jamaican-born Michael Marley, a wealthy developer (and a second cousin of reggae musician Bob Marley), and Beverly, his winemaking spouse, could indulge their new-found passion for growing grapes anywhere in the world but chose to establish Marley Farm Winery on their farm in the Saanich Peninsula. They are among the growing number of gentlemen farmers who have recently established small Saanich vineyards, either to sell grapes to established wineries or to open new wineries in the future.

It has been trial and error for many of these wine growers, separated by half a day's drive plus

OPPOSITE: **SPRINGTIME AT GLENTERRA VINEYARDS NEAR COBBLE HILL.**

a ferry ride from their more experienced peers in the Okanagan. Michael Betts, a former builder of yachts and experimental aircraft who opened Chalet Estate Vineyard in 2001 with partner Linda Plimley, was mentored initially by John Kelly, who apologized, Betts recalls, "that his knowledge was twenty minutes earlier than mine." Even Busnardo, who

CHALET ESTATE'S MICHAEL BETTS HAND-LABELLING BOTTLES OF WINE.

grew up amid grape vines in Italy and who planted in the Okanagan in 1968, put some vines in a wet spot in his Cowichan Valley vineyard, only to see them fail. Dosman planted his gently sloping 10-acre (four-hectare) vineyard in 1994 with 30 varieties, including "goofy stuff, stuff that you try." He began correcting his planting errors three years later. A purist, he makes wine only from estate-grown grapes and, unlike many other island winemakers, eschews purchased Okanagan grapes. As a result, Alderlea Vineyards began generating a living for Dosman and his family only after its 2001 vintage reached the market. Dosman is the model practical and hard-headed winemaker (so frugal that in 2002 the winery had just one line for both telephone and fax).

Others are simply enraptured by the wine grower's lifestyle. In 1995, when he was 72, Lorne Tomalty, a retired senior administrator in British Columbia's civil service, began planting for his proposed Malahat Estate Vineyard. The steep, rock-strewn four-acre (1.6-hectare) southeast-facing vineyard is very near the Malahat Summit, north of Victoria. At an elevation of 630 feet (192 meters), it is one of the highest vineyards on the island. "I think wine is a great thing," the vigorous Tomalty proclaims, explaining why he embarked on wine growing at his age. "There is something about being in the middle of a vineyard," Marley Farm's Michael Marley agrees, still speaking with the lilt of the Caribbean a quarter of a century after arriving in Victoria. "It's a marvellous experience."

Perhaps because the industry is so driven by inherent romance, wine growing has burgeoned here in the past decade. Fraser Smith, who formed the Vancouver Island Grape Growers Association in 1997, believes that as many as 50 wineries will thrive here one day. This is not wishful thinking since nearly 20 wineries have opened since 1990. They are making wines of character: fresh and crisp whites and reds, frequently with Pinot Noir or with the ubiquitous Ortega. Pinot Noir is challenging because the ripe bunches are susceptible to mildew in anything but a dry climate. That problem has confronted Sonny Vickery, owner of Vicori Estate Winery on the Saanich Peninsula, which was included on industry winery lists prematurely. The 4,000 vines in the Vicori vineyard also include Ortega and Müller-Thurgau; the latter also is notoriously prone to mildew.

Several producers, including Chateau Wolff at Nanaimo, Venturi-Schulze Vineyards and Alderlea near Duncan and Newton Ridge in the Saanich Peninsula, compete aggressively to make the island's best Pinot Noir. Harry von Wolff, who was born in Riga, the capital of Latvia, in 1934, is a man of eclectic interests who claims to have worked in perhaps sixty different jobs from ranching to clothing retailing before establishing Vancouver Island's most northern winery in 1997. His love of wine, notably wines of Burgundy, developed during his studies of hotel management in Switzerland. Today, the warmest spots in his vineyard are reserved for Pinot Noir, which he makes in a full and muscular

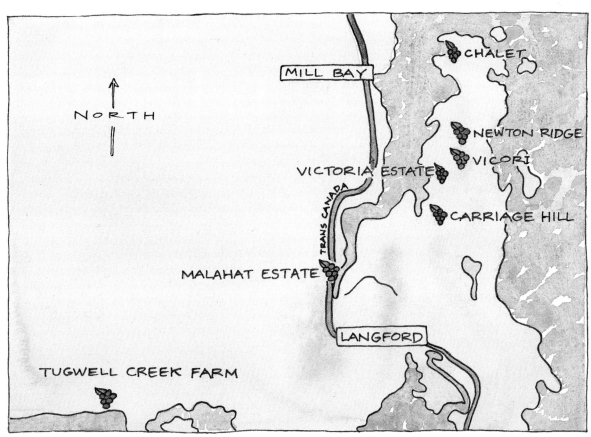

style, and Chardonnay. The vineyard, on the western edge of Nanaimo, is nestled against a height of land which retains the heat from the sunlight falling on his southwestern slopes during the day.

At Yellow Point Vineyards near Ladysmith, Chris Pinnock also planted Pinot Noir, along with Chardonnay and Pinot Blanc, on his century-old farm just behind the Nanaimo Collishaw Airport. "I like a full-bodied red or a well-oaked Chardonnay," Pinnock says. "I think we have the right conditions to ripen the grapes and produce a distinctive, complex wine." Born in Yorkshire, Pinnock became a telecommunications technician in the British Army. That skill led him to Saudi Arabia where he maintained the radar and flight data systems at Riyadh's international airport. "Alcohol is illegal in Saudi Arabia, so everyone makes his own wine and beer," says Pinnock, an amateur winemaker since adolescence. "Most people use grape juice. By experimenting with temperature, sugar content, fermentation time and other ingredients, one can make a pretty decent wine." Pinnock and Johanne, his Edmonton-born spouse, moved to Vancouver Island after completing work contracts in Saudi Arabia. "I think touring the island wineries sparked the idea for our own winery," he says. Their winery is under development.

Of all the new wine growers on the island, few have prepared as assiduously as Andrew Johnston, the owner of Averill Creek Vineyards and Winery at Duncan, which expects to release wine in 2006. Johnston, who was born in England in 1947 and came to Canada in 1973, is a wine-loving Edmonton doctor. He and a partner owned 24 walk-in medical clinics in Edmonton and Calgary. But throughout

OPPOSITE: **WILD SPRING FLOWERS ENHANCE THE CHARM OF VIGNETI ZANATTA'S VINOTECA RESTAURANT.**

PINOT NOIR PRODUCES SOME OF VANCOUVER ISLAND'S BEST RED WINES.

THE CAREFULLY GROOMED VINEYARD AT THE BLUE GROUSE WINERY.

his medical career, Johnston planned a second career as a winemaker. He has volunteered at capable wineries, beginning in 1998 with a winery in Tuscany owned by Umberto Menghi, the Italian-born Vancouver restaurateur. Menghi's wines include a super-Tuscan red called Bambolo. "I did the cap management on that wine in 1998," Johnston says. He has done subsequent vintages in Australia, France and New Zealand (two in the latter country). While doing this, he searched for vineyard sites, looking at properties in France, New Zealand and Tuscany until his daughters persuaded him to remain in Canada. He acquired a 46-acre (18.6-hectare) property north of Duncan, on the fairly steep southward slope of Mount Prevost with excellent exposure. He will exploit the terrain to build a winery employing gravity to move the wine.

"When you walk to the top of the vineyard and look down, it is just stunning," Johnston marvels. He planted 14,400 vines in 2002, including four clones of Pinot Noir, along with Pinot Gris, Gewürztraminer and Merlot. "It's an expensive place to start a vineyard," he says of Vancouver Island. "It's an expensive place in which to make wine. But for me, it is lifestyle as well as practicality. I enjoy living there. After 25 years in Edmonton, I've had enough of 25-below winters."

With such a robust revival of winemaking on Vancouver Island, commercial winemaking in British Columbia has returned to where it began about 1920, with loganberry wines made on southern Vancouver Island. Over time, berry wines lost favour to grape wines. Growers' Wine Company, the only original Vancouver Island winery to survive more than a few years (because it switched to wines made with Okanagan or California-grown grapes) moved in 1978 to a new winery in Surrey on the Lower Mainland. In turn, the Surrey winery,

VANCOUVER ISLAND VINEYARDS

JOE BUSNARDO
DIVINO ESTATE WINERY

HARRY VON WOLFF
CHATEAU WOLFF

ROGER AND NANCY DOSMAN
ALDERLEA VINEYARDS

WAYNE AND HELENA ULRICH
CHERRY POINT VINEYARDS

JOHN KELLY
GLENTERRA VINEYARDS

**MICHAEL BETTS AND
LINDA PLIMLEY**
CHALET ESTATE VINEYARD

**MICHELE SCHULZE, MARILYN
SCHULZE AND GIORDANO VENTURI**
VENTURI-SCHULZE VINEYARDS

LORETTA ZANATTA AND JIM MOODY
VIGNETI ZANATTA

FRASER SMITH
VICTORIA ESTATE WINERY

DAVID GODFREY
GODFREY-BROWNELL VINEYARDS

EVANGELINE AND HANS KILTZ
BLUE GROUSE VINEYARDS

which had operated as Jordan & Ste-Michelle Cellars, closed in 1990, just as winemaking was reviving on Vancouver Island.

Taking advantage of the gentle climate of the Saanich Peninsula, the early farmers planted extensive acreages of berry bushes, most notably the succulent loganberry. This berry had been developed in Oregon by a horticulturist named Logan who crossed the wild blackberry with raspberry. Faced with a huge loganberry surplus, the Saanich farmers formed the Growers' Wine Company in 1923 "to create a market for the Loganberry, which at that time was practically unsaleable," as Growers' recounted in a brochure a few years later. Their timing was good. British Columbia in 1921 ended four years of Prohibition and the demand for wine was met with berries because the Okanagan's vineyards only began producing in commercial quantities in the late 1920s. The brochure recorded that the wine, produced in a winery on Quadra Street in downtown Victoria, was made "from the juice of freshly picked, ripe Loganberries and cane sugar only [and] is of the Port class, having a proof percentage of 28°." By 1926 Growers' was buying more than 80 percent of the island's loganberries. "The development of the wine industry by the Growers' Wine Company has meant the salvation of the Loganberry industry," the brochure boasted. These berry wines continued to be sold into the 1960s, when a plant virus decimated the loganberry fields.

Growers' was not alone. Stephen Slinger had a small berry winery at Chemainus and may have been bootlegging wine before Prohibition ended. A berry winery was established at Richmond on the mainland and another Saanich producer, Brentwood Products, opened in 1927 in Brentwood Bay, recruiting Slinger to make the wines. Brentwood

NETS PROTECT THE MATURING GRAPES AT BLUE GROUSE FROM VORACIOUS FLOCKS OF BIRDS.

soon renamed itself Victoria Wineries and converted an old Victoria hotel to winemaking. The company is on record for offering a generous $100 a ton in 1930 for Okanagan grapes. The shareholders of Victoria Wineries learned that it is easier to make wine than to sell it: by the end of 1930, the winery had 160,000 gallons (727,360 litres) of wine in storage

ENVIRONMENTALLY-SOUND VINEYARDS AT CHERRY POINT WINERY STRIVE FOR HARMONY WITH NATURE.

but sales in 1931 averaged only about 3,000 gallons (13,640 litres) a month. "A suggestion for Xmas," a winery executive implored shareholders late that year. "Buy a carton of six…at $3.50, and distribute among your friends." By the mid-1930s, Growers' had taken over all of the island wineries and was buying Okanagan grapes, there being no significant vineyards on the island at the time. In another example of the industry coming full circle, a new Victoria Estate Winery, this time with a newly planted vineyard of its own, opened in 2002, not far from world-renowned Butchart Gardens in Saanich. Victoria Estate, however, debuted with wines made from Okanagan grapes and it plans to keep supplementing island-grown grapes with Okanagan fruit.

Berry wines made a small comeback on the island in 2001 when Cherry Point Vineyards released a port-style wine from wild blackberries. For several years, First Nations pickers had been selling their berries to consumers in Cherry Point's parking lot, with the winery's owners making berry pies and trial lots of wine with any unsold berries. In 2001, the winery asked for enough berries for a serious volume of wine — and received 5,000 pounds (2,268 kilograms). The wine was ready to drink before Christmas and sold out by Easter and is destined to be made annually. "There is no reasonable limit to the amount of berries we can buy," Cherry Point owner Wayne Ulrich says.

Merna Moffat agrees. "We figure there are over 100,000 pounds easily on the island," she says. With husband, Walter, she is proposing to establish the Honeymoon Bay Wild Blackberry Winery at the community of Honeymoon Bay on Cowichan Lake, west of Duncan. The Moffats — she is a laboratory technician and he is an accountant — retired to Honeymoon Bay from Prince George in the early 1990s. With the decline of forestry jobs, the Moffats believe that a winery would help generate tourism in the community. "We are sitting on a beautiful lake which is virtually undeveloped," she says. By 2002, they had produced trial lots of wine in styles from nearly dry to sweet and were looking for investors to move the venture forward.

Merridale Ciderworks, British Columbia's first estate cidery, produces a range of classic unpasteurized English ciders from the fruit of 3,500 cider apple trees in its 13-acre (5.26-hectare) orchard near Cobble Hill in the Cowichan Valley. As well, it has inspired several other growers in the valley to plant cider apples, which are quite different from eating apples. "What you get from a cider apple is tannin and acids, much like a wine grape," says Janet Docherty, a Victoria business person who acquired Merridale in 2000 with her husband, Rick Pipes, a Victoria lawyer and now a cider maker as well. Merridale, which opened in 1992, was founded by Albert Piggott, a Scots teacher who brought a palate for cider with him when he immigrated to Canada in 1954. "Al made really good cider," Docherty says. The rustic facility has been upgraded to become an appealing destination on Cowichan Valley wine

THIS RETIRED WINE VAT, NEARLY A CENTURY OLD, NOW LANDMARKS CHERRY POINT WINERY.

tours. While not a wine producer, Merridale joined the Vancouver Island Vintners Association and Docherty even became the association's president in 2002. Merridale continues to make traditional English ciders, full-flavoured artisanal beverages quite different from the mainstream apple ciders made from familiar dessert apples. "A cider made from an eating apple is like a wine made from Thompson seedless grapes," Docherty maintains.

The first modern commercial vineyard, now 30 acres (12 hectares) in size, began in 1970 with an experimental planting by Dennis Zanatta, on a former dairy farm just south of Duncan. A wine-loving immigrant from northern Italy, he was confident that wine grapes could grow in the Cowichan Valley. The valley, shielded from the Pacific Ocean's storms by nearby mountains, has a long growing season. Dry enough that vineyards usually are irrigated (as they are on the surprisingly arid Gulf Islands), it is frost-free from mid-April until late October or even mid-November, long enough for many grape varieties to mature.

Zanatta's trials caught the attention of the provincial government, which used his vineyard to assess about 100 different varieties. The Duncan Project, as this test was called, ran from 1983 to 1990, and had identified Ortega, Auxerrois and Pinot Gris as promising varieties when the funding ran out prematurely, long before reliable conclusions were reached. The final report on the project observed that it would take 20 to 30 years to determine whether a region is suitable for growing wine grapes.

Paralleling this project, a private vineyard trial was begun in 1983 at nearby Cobble Hill by John Harper, an experienced grape grower who previously operated a vineyard in the Fraser Valley near Langley. Harper had some bad luck with partners

and had to start over at another Cobble Hill property. He died in 2001, by which time his judgment in picking vineyard locations was proven. German-born Hans Kiltz's thriving Blue Grouse Vineyards and Winery, which opened in 1993, grows Pinot Gris, Ortega and Gamay, among other varieties, on the sun-bathed slope that was Harper's first island vineyard. The second Harper vineyard was purchased in 1998 by John Kelly and partner Ruth Luxton, who named it Glenterra Vineyards. Harper's legacy to Glenterra included a trial plot with 40 different grape varieties on a single acre and a greenhouse with a lemon tree and a vine of Dornfelder, an early-maturing German red.

The sporadic grape trials did not discourage the determined. Dennis Zanatta sent his daughter, Loretta, to a wine school in Italy. When she returned with her master's degree and a sure hand at making sparkling wine, they opened Vancouver Island's first modern winery, Vigneti Zanatta, in 1992. This neighbourhood just south of Duncan is known as the Glenora district. The name is honoured by the winery's flagship sparkler, Glenora Fantasia, a crisp but fruity wine made from the Cayuga grape. This is a white grape with Muscat flavour tones that was developed in New York State precisely for cool-climate grape growing like that found in the Cowichan Valley. No other grower has taken a chance on Cayuga, but for Zanatta the gamble paid off.

An Italian heritage also informs Venturi-Schulze Vineyards, which opened in 1993. Giordano Venturi's infectious passion for wine is shared by Australian-born Marilyn Schulze, his spouse. Both were science teachers until 1988 when they began turning a century-old farm along the highway near Cobble Hill into a meticulously groomed vineyard. The wines, often highly original blends, have achieved cult status. They are snapped up within days of being announced to those fortunate enough to be on the mailing list; Venturi-Schulze almost never opens its tasting room to the public. The winery is equally renowned for its balsamic vinegar, made from wine that is boiled to reduce its volume and then aged for years in chestnut and other exotic barrels. This exquisite vinegar is a tradition

FRIENDLY VINEYARD DOGS ARE A WINERY TRADITION.

OPPOSITE: **COMPACT WINERY SERVES GODFREY-BROWNELL VINEYARDS.**

in Modena, the Italian city near which Venturi grew up; it has now become a tradition at his tiny winery on Vancouver Island.

At the premature conclusion of the Duncan Project, more than 30 of the test varieties were moved to the Cherry Point vineyard that Wayne and Helena Ulrich, attracted by an optimistic view of the wine-growing lifestyle, established in 1990 on a 35-acre (14-hectare) former mink ranch near Cobble Hill. Raised on a Saskatchewan farm and trained as an engineer, Wayne Ulrich was a comfortable federal farm bureaucrat processing grants for Okanagan viticulture. When this stirred an interest in wine, he and Helena, a lamp store owner in Victoria, scoured Vancouver Island, thermometer in hand, to find a suitable location. They planted Pinot Blanc, Pinot Gris and Gewürztraminer, which proved successful, and Müller-Thurgau, which failed. They also transplanted some Duncan Project vines into their own test plot. The star in the trial was Agria, now being adopted by other island wine growers. Developed in 1965 in Hungary, early-ripening Agria produces an inky black and quite substantial wine, with an occasional tannic muscularity rare among cool-climate reds. "We've been excited about it since we first picked the fruit," Ulrich says. Cherry Point began selling wines in 1994; five years later, it released the world's first varietal table wine from Agria, a mere 23 cases from the 1998 vintage. By the 2001 vintage, Cherry Point's annual production included 200 cases of Agria, sold primarily at the tasting room. With its summertime Sunday picnics, music concerts on the lawn and sheep in the meadow, Cherry Point has become one of the Cowichan Valley's destination wineries.

Wayne and Helena Ulrich, who had not made a single bottle of wine prior to launching Cherry Point, now have planted most of their property and intend to increase production to a maximum of 4,500 cases, making it one of the island's largest. Cherry Point could even phase out its use of Okanagan grapes. However, after experiencing 10 vintages on the island, Ulrich can also make the case against relying exclusively on island-grown grapes. His first year as a grape grower, 1992, was a "wonderful year" but birds — a major problem for island vineyards — devoured the crop because the vines had not been netted. The next two vintages were good, followed by two mediocre vintages and then a 1997 ruined by a disastrously cool summer. The 1998 vintage was "golden" in quality, followed by a cool and damp 1999, a mediocre 2000 and a 2001 saved only by a prolonged warm and dry autumn. "Not an awful lot of very good years," Ulrich says. "I don't know if we've come up with anything but plant and hope. Outside of crop thinning, there are no techniques to improve ripening."

That is not quite so. There is a practice called tenting, employed by Roger Dosman at Alderlea with a limited number of vines and by Giordano Venturi at Venturi-Schulze over much of the vineyard. (Fraser Valley vineyards also have begun to use this technique.) Polyethylene sheets are draped over the vines in spring to accelerate growth through a greenhouse effect, gaining perhaps two weeks of additional maturity. As a result, Alderlea's 2001 Merlot was a riper wine than the 2000. Tenting is not needed for every variety but may be the only solution for late-maturing grapes. David Godfrey fully expects that the Cabernet Sauvignon planted at Godfrey-Brownell Vineyards near Duncan will need to be tented to ripen well.

Godfrey is among those who believe that the first duty of wine is to be red; half of Godfrey-Brownell's 20-acre (eight-hectare) vineyard, established in 1999, is committed to red varieties, with the aim being to produce "a full-bodied island red." The winery also makes other reds from Okanagan grapes. Godfrey would like to source both Sangiovese and Nebbiolo, two classic Italian red varieties, so that he might create the local equivalents of Montepulciano and Barolo, an ambition that marks him as another romantic. Born in Winnipeg in 1938, he grew up near Toronto in a neighbourhood of immigrant Italians who made their own wine every autumn and gave him his first pointers. Godfrey pursued an academic career, teaching literature at the University of Toronto and then at the University of Victoria after moving to Vancouver Island in 1976. A man of boundless interests, Godfrey once built his own computer and later wrote several books on technology before launching, with his wife, Ellen, an Internet service provider in Victoria.

Godfrey's grandparents retired in the 1930s, moving from the Prairies to a small property on Cadboro Bay where they grew loganberries and a few grapes. The property remains in the family, complete with a grizzled Concord grapevine, as a small organic farm. In 1993 Godfrey, deciding to grow his own grapes, began researching vineyard sites and economics with academic thoroughness. "It became clear to me that you could not break even on less than 12 acres of grapes," he learned. "Eric von Krosigk, who was my consultant, said 20." With nothing large enough or cheap enough on the

Saanich Peninsula, Godfrey took five years to locate a well-drained, gravelly property in the Cowichan Valley, immediately south of the Zanatta vineyards. Searching the title, Godfrey discovered that the land had been homesteaded in 1886 by one Amos Aaron Brownell, his grandmother's second cousin. "It was meant to be," he says, "so we felt we had to put Brownell in the name of the winery."

When planting, Godfrey benefitted from the experience of others, initially putting in two "insurance" varieties, Bacchus and Maréchal Foch, and commercial mainstream grapes including Pinot Gris, Pinot Noir, Gamay and Chardonnay. When the vineyard was expanded by a third in 2002, the new varieties included Cabernet Sauvignon, Merlot, Agria, Lemberger and Dunkelfelder. "Out of all of that, we'll get a full-bodied island red," he hopes.

The development of vineyards at Zanatta and Godfrey-Brownell has led to a third emerging winery in the Glenora area near Duncan. Echo Valley Vineyards has been established by Albert Brennink, 78, and his son Edward, in a warm valley about one and three quarter miles (three kilometres) west of the Zanatta winery. The elder Brennink is a Dutch architect who had a thriving practice restoring churches, schools and homes in postwar Europe. Before he retired in 1977 and brought his family to Canada, Brennink had lived near Geneva in Switzerland and acquired an interest in wine there. Edward, in the meantime, pursued a career as an agriculturist. On the farm near Duncan, the Brenninks cultivated Scottish Highland cattle, a docile breed, for beef. When they concluded that a portion of their property was suitable for grapes, they planted a one-acre (.4-hectare) vineyard with

NETS PROTECT THE TINY NEWTON RIDGE VINEYARD AGAINST BIRDS.

40 experimental varieties. The best varieties will ultimately be planted on about 20 acres (eight hectares). To get an early start, Echo Valley engaged consultant Eric von Krosigk to make wines for the winery in the Okanagan in 2001. The winery aims at a style in what Edward Brennink calls the "northern France direction" of Alsace and Burgundy.

Almost every island vineyard grows Ortega. A cross between Müller-Thurgau and Siegerrebe, the Ortega was developed in Germany to be a cool-climate variety, maturing early but achieving relatively high sugars. It is a bread-and-butter white, yielding fruity but dry wines with spicy aromas of peaches or apricots. It has, however, been a hard sell to consumers, as former Newton Ridge proprietor Peter Sou discovered. Born in Holland and now a retired Bechtel Corporation engineering executive, Sou acquired Newton Ridge in 1996 after the three-acre (1.2-hectare) vineyard had been planted to Pinot Gris, Pinot Blanc, Pinot Noir — and Ortega. One of Canada's smallest wineries with an annual production of about 350 cases, Newton Ridge began selling its wines from the 1998 vintage. "When I tried to sell my Ortega to restaurants, no one wanted it," Sou recounted. Then Sinclair Philip, owner of the elegant Sooke Harbour House, a restaurant with a stupendous wine cellar, came to dinner at Newton Ridge. After tasting a tank sample of the 1999 Ortega, he bought Sou's entire production. "Now everybody wants it," the winery owner says. "This might put Ortega on the map." That task must be completed by someone else because the non-drinking Sou sold Newton Ridge in 2002 to Tom Shoults and Nicki Suvan.

The winery farthest from the Vancouver Island wine trail, or perhaps at the northwestern terminus,

VINEYARD AT CHATEAU WOLFF IN NANAIMO, ONE OF THE MOST NORTHERN ON VANCOUVER ISLAND.

is Port Alberni's Chase & Warren Estate Vineyards. While precipitation is high in the Alberni Valley, the weather patterns are such that summer and autumn are hot and fairly dry. The city, after all, is in the middle of Vancouver Island and the Pacific is 30 miles (50 kilometres) away at the western end of the Alberni Inlet. Orchards once flourished at Port Alberni, growing apples, plums, cherries and raspberries. (Jan Peterson, a local historian, recounted in her book *The Albernis 1860–1922* that about 1885 a member of a First Nations band developed an orchard. "Several hundred fruit trees were planted and eighty of them soon produced fruit.") The winery is the project of elementary school teacher Vaughan Chase and his brother-in-law, Ron Crema. A Victoria native, Chase lives on a southwest-facing escarpment overlooking the valley. The first intimation that he could grow wines here came when a handful of decorative Gewürztraminer vines on his property did well and yielded pleasant wine. Urged on by Crema, who is of Italian background, Chase in 1996 began by planting five acres (two hectares) of grapes, including Bacchus, Siegerrebe, Pinot Gris and Agria. "There is a lot more potential in the valley for grape growing," Chase believes. The Beaufort Range — the mountains defining the valley's northern boundary — has ideal southwest slopes of well-drained gravel.

Shortly after Chase began his vineyard, neighbour Evan McLellan started a nine-acre (3.6-hectare) vineyard, primarily with Schönburger, Madeleine Angevine, Pinot Gris and Reichensteiner. The owner of a construction company, McLellan was raising livestock on a hobby farm until he developed an allergy to hay. He settled on grapes as an agricultural alternative. A native of Port Alberni, where he was born in 1955, McLellan has long-term plans for a winery.

The Tugwell Creek Farm at Sooke, on the west coast of Vancouver Island, breaks new ground. This is a honey farm where Robert Liptrot and Dana LeComte are opening British Columbia's first meadery in 2003, operating under a fruit winery licence because regulations for mead production do not exist. Now 47, Vancouver native Liptrot, who has a master's degree in entomology, has been keeping bees since he was seven. He was not much older when he began fermenting mead, an ancient alcoholic beverage that was made in pre-Roman Britain. "It survived as a drink until fairly recently, being made in country houses, but at the present time it is rather a curiosity and not much of a commercial drink, although it is commercially made," Pamela Vandyke Price wrote in the 1980 edition of *Dictionary of Wines and Spirits*. Liptrot's research uncovered only a handful of mead producers in New Zealand, the United States and Quebec.

Liptrot and LeComte moved to Vancouver Island in the mid-1990s to raise bees there because the island was free of a mite that has devastated bee colonies elsewhere. Sadly, the mite arrived soon after they did but, with effective disease management, they have succeeded in building a business. They have about 100 hives in mountain clearings on southern Vancouver Island, where the bees produce pure honey — averaging about 110 pounds (50 kilograms) per hive a season — from the nectar of fireweed and salal. Most of the honey is sold but about a third is turned into mead. Liptrot has started out with three basic styles. The driest, not unlike a sherry, is vintage mead, bottled only after spending four years in oak barrels and crafted to age in bottle another 15 or more years. A mead style called melomel (from an old Anglo-Saxon word) refers to mead mixed with berry juice; in Tugwell Farm's case, the juice includes farm-grown English gooseberries, loganberries and marionberries. The sweetest style, called sack, is a dessert style somewhat analogous to icewine. "Mead is one of those end products which typifies beekeeping at its pinnacle," Liptrot believes.

THE WINERIES

Alderlea Vineyards
1751 Stamps Road, RR 1, Duncan, BC, V9L 5W2
Telephone: 250 746-7122

Averill Creek Winery
6552 North Road, Duncan, BC, V9L 6K9
Telephone: 250 715-7379
www.averillcreek.ca

Blue Grouse Vineyards
4365 Blue Grouse Road, Duncan, BC, V9L 6M3
Telephone: 250 743-3834
www.bluegrousevineyards.com

Chalet Estate Vineyard
11195 Chalet Road, North Saanich, BC, V8L 5M1
Telephone: 250 656-2552
www.chaletestatevineyard.ca

Chase & Warren Estate Winery
6253 Drinkwater Road, Port Alberni, BC, V9Y 8H9
Telephone: 250 724-4906

Chateau Wolff
2534 Maxey Road, Nanaimo, BC, V9S 5V6
Telephone: 250 753-9669

Cherry Point Vineyards
840 Cherry Point Road, RR 3, Cobble Hill, BC, V0R 1L0
Telephone: 250 743-1272
www.cherrypointvineyards.com

Church & State Wines
(formerly Victoria Estate)
1445 Benvenuto Avenue, Brentwood Bay, BC, V8M 1R3
Telephone: 250 652-9385
www.churchandstatewines.com

Divino Estate Winery
1500 Freeman, Cobble Hill, BC, V0R 1L0
Telephone: 250 743-2311

Echo Valley Vineyards
4651 Waters Road, PO Box 816, Duncan, BC, V9L 3Y2
Telephone: 250 748-1470
www.echovalley-vineyards.com

Glenterra Vineyards
3897 Cobble Hill Road, Cobble Hill, BC, V0R 1L0
Telephone: 250 743-2330
E-Mail: glenterravineyards@shaw.ca

Godfrey-Brownell Vineyards
4911 Marshall Road, Duncan, BC, V9L 6T3
Telephone: 250 715-0504
www.gbvineyards.com

Honeymoon Bay Wild Blackberry Winery
Honeymoon Bay, BC, V0R 1Y0
Telephone: 250 749-4681

Malahat Estate Vineyard
1197 Aspen Road, Malahat, BC, V0R 2L0
Telephone: 250 474-5129

Marley Farm Winery
1831 D Mount Newton Crossroad, Saanichton, BC, V8M 1L1
Telephone: 250 652-8650
E-Mail: carriage.hill@shaw.ca

Merridale Ciderworks
1230 Merridale Road, RR1, Cobble Hill, BC, V0R 1L0
Telephone: 250 743-4293
www.merridalecider.com

VANCOUVER ISLAND VINEYARDS

THE WINERIES

Newton Ridge Vineyards
1595 Newton Heights Road, Saanichton, BC,
V8M 1T6
Telephone: 250 652-8810

Tugwell Creek Farm
8750 West Coast Road, Sooke, BC, V0S 1N0
Telephone: 250 642-1956
www.tugwellcreekfarm.com

Venturi-Schulze Vineyards
4235 Trans Canada Hwy, Cobble Hill, BC, V0R 1L0
Telephone: 250 743-5630
www.venturischulze.com

Vicori Winery
1890 Haldon Road, Saanichton, BC, V8M 1T6
Telephone: 250 652-4820
E-Mail: svickery@shaw.ca

Vigneti Zanatta
5039 Marshall Road, RR 3, Duncan, BC, V9L 6S3
Telephone: 250 748-2338
www.zanatta.ca

** Yellow Point Vineyards
13386 Cedar Road, Ladysmith, BC, V9G 1H6

** UNDER DEVELOPMENT

GULF ISLANDS VINEYARDS

"MY VIEW IS IF ANYBODY WANTS TO START A WINERY [ON THE GULF ISLANDS], THAT'S FINE — AS LONG AS THEY MAKE GOOD WINE."
—MARCEL MERCIER
GARRY OAKS ESTATE WINERY

It is propitious, Marcel Mercier suggests, that he can see the Burgoyne Valley from the highest corner of the Garry Oaks Estate Winery that he and partner Elaine Kozak have built on Salt Spring Island. The valley name, he says, may be a corruption of *Bourgogne*, one of the great French wine regions. The wines of Burgundy are a quality benchmark against which Garry Oaks and other new wineries on the Gulf Islands aspire to measure themselves.

Mercier's aspirations are admirable even if the facts spoil the story. Like many other coastal place names, Burgoyne Road, which leads to Burgoyne Bay on the west side of Salt Spring, is named for a British naval officer, Hugh Burgoyne. Salt Spring itself was named Admiral Island by the naval surveyors but the residents stubbornly retained the older name given by Hudson's Bay traders, inspired by the island's brine springs. Though the Gulf Islands have little in common with the region, Burgundy has stirred several of the island vintners. Mayne, Pender and Salt Spring once had extensive orchards; at the Garry Oaks Vineyard, Mercier and Kozak have retained several apple trees to preserve heritage varieties such as the King apple. Until they were dispossessed in 1942, tomato-growing farmers of Japanese origin comprised a third of Mayne Island's population. On Saturna, where sheep and other livestock are still raised,

OPPOSITE: **NEWLY PLANTED GRAPES IN THE VINEYARD ON SATURNA ISLAND.**

an annual lamb barbecue on the beach once was a highlight of island life. Perhaps the least agricultural of the islands was Galiano, because it is very dry and was formerly a timber preserve owned by MacMillan Bloedel.

Even with agricultural history to draw on, the wine growers face challenges unique to the islands, from finding enough irrigation water to ripening grapes in vineyards chilled by cool sea breezes. "Salt Spring really has a limited number of suitable sites just because the mountains all face the wrong way," Garry Oaks co-proprietor Elaine Kozak says. "The accessible slopes on the mountains tend to face north. The sharp sides are facing south and the more gentle slopes are facing north." Because this is a new wine region, vineyard and winery supplies are not readily available. The owners of the three wineries being developed on Salt Spring share such equipment, wherever practical. On the sparsely populated islands, there are too few vineyard workers and too few residents to drink the wine that is made. None of the wine growers has prior experience at growing grapes and seldom much experience at making wine. On the other hand, all are highly skilled and educated. For example, Arlene and Norman Kasting, who own Overbury Farm on Thetis Island, both have doctorates. With such brain power, every problem finds a creative solution. The other Thetis Island wine growers, Colin Sparkes and his wife, Carola Daffner, are international software consultants turned resort developers. Sparkes plans to foster sales through island-hopping wine tours. Given the climate and the grape varieties, the island wines should be the ultimate crisp, fresh expression of cool-climate wines, a fine match to the fresh oysters that Sparkes harvests along the shore of his Cedar Beach resort.

The pioneers have been The Vineyard at Bowen Island and Saturna Island Vineyards. A short ferry ride from West Vancouver, Bowen Island is both a bedroom and a playground for city folks. Lary Waldman and his wife, Elena, who have lived on the island since 1978, began planting a six-acre (2.4-hectare) vineyard in 1994 at the same time as they developed a guest house and small conference centre. Waldman released his first wines in 1999, made from varieties including Pinot Blanc, Bacchus and Pinot Noir. They were sold almost entirely on the island. However, the winery closed in 2002.

Saturna Island's vineyard is 10 times as large and the wines are sold widely. This is a matter of commercial reality, given the winery's substantial production. There are perhaps 300 permanent residents on Saturna. Many tourists visit the island during the season, some to stay at Saturna Island Lodge, a seven-bedroom country inn with a dining room. Yet because the lodge and the vineyard have the same owner, Saturna's wines cannot be served in the lodge's fine restaurant. An ancient British Columbia regulation, the so-called tied house rule, prohibits a brewer, distiller or vintner from selling his own products in any restaurant he owns other than one at the premises where the beverage is made. The rule dates from the 1920s when brewers or distillers actually owned pubs and hotels and has never been repealed, even if the relevance is dubious today. On Saturna Island, the lodge is out of bounds to the vineyard's wines because it is four and a quarter miles (seven kilometres) from the winery. Regulators refused to waive the rule.

The southernmost of the Gulf Islands, Saturna takes its name from the ship *Saturnina* with which the Spanish explored these waters in 1791. There was little development beyond sheep farming and a small resort until 1990, when a group of investors headed by Larry Page, a Vancouver securities lawyer, purchased substantial acreage on the southwest side of the island. The shoreline property was subdivided for elegant housing. At the suggestion of a French restaurateur in Vancouver, Page planted vines on an elevated bench of land away from the shore, against a towering, west-facing sandstone cliff that captures the sun's heat. Beginning in 1996, four large vineyard blocks have been planted, totalling 60 acres (24.3 hectares). The varieties include Pinot Noir, Chardonnay, Pinot Gris, Gewürztraminer, Muscat and Merlot. The first wines were released in 1999, beginning with a Sémillon made from grapes purchased in the Okanagan. Saturna Island Vineyards is among a number of wineries

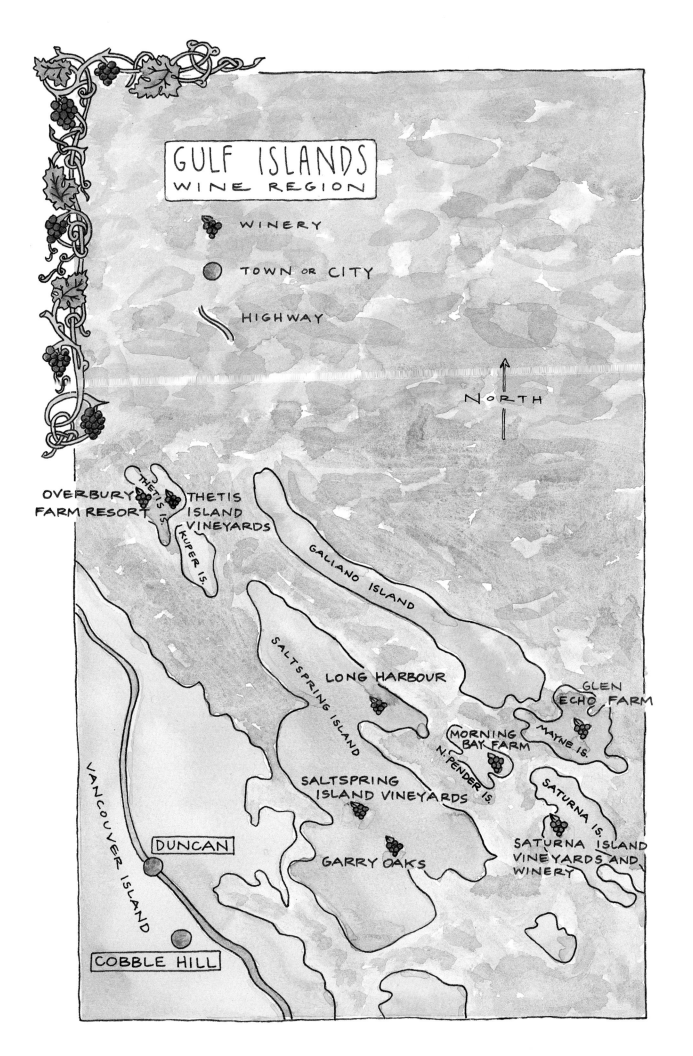

on the Gulf Islands and Vancouver Island that also purchase Okanagan grapes. This has been contentious with those other producers that use only grapes grown in their vineyards or their region. Yet a large winery like Saturna Island (8,000 cases in 2000), which wants to offer a full range of wines, has little practical option. Varieties, notably the Bordeaux reds, which ripen well in the Okanagan, struggle to mature on the Gulf Islands.

The Morning Bay Farm vineyard on North Pender Island, with three acres (1.2 hectares) planted initially in 2002, is directly across Plumper Sound from Saturna Island Vineyards. (The *Plumper* was a British survey vessel plying island waters from 1857 to 1861. Both Daniel Pender and Richard Mayne were officers on the ship and gave their names to islands.) Morning Bay Farm is the project of Keith Watt, a radio journalist and media instructor at Capilano College in North Vancouver, and his partner, Janet Clothier, a legal administrator. "I really inaugurated this project on my fiftieth birthday," says Watt, who was born in Winnipeg in 1951. He and Clothier purchased the rugged 25-acre (10-hectare) property in 1992. They were inspired to plant grapes by Saturna's burgeoning vineyard. "I come to agriculture with my eyes open," he says. Watt once worked for an impecunious Ontario farmer whose barn had burned. He helped the farmer salvage and straighten used nails for its reconstruction. As a broadcaster for the Canadian Broadcasting Corporation in Edmonton, Watt produced farm shows and won awards for his agricultural documentaries. In 2001, he took a one-year sabbatical from his college job, volunteering at Saturna Island to experience the year in a vineyard before preparing his property for vines. "Our site is a really difficult site," he says, comparing some of the steep southward-facing slopes to the notoriously steep vineyards of the Mosel. Watt recognizes that the property will be difficult to farm, but it has the high scenic values appropriate for a bed and breakfast and a winery that can take advantage of gravity. Morning Bay, however, is not intended to rival Saturna Island in size. "I see Morning Bay as making handmade wines in small lots going to friends."

Watt, who prefers the wines of Burgundy, has planted both Pinot Noir and Pinot Gris. He decided against Chardonnay because there is a glut of wines from this variety. On the advice of Saturna Island's Eric von Krosigk, Watt also planted Schönburger, 800 Gewürztraminer vines and 400 Riesling vines. Watt suggests that with Riesling "if you only get just 15° Brix, you can still make a pretty decent wine. Looking at its success in the Mosel, which seems to be the temperature profile most similar to what we're facing, it looks like Riesling might be a really good grape here." On one of the property's coolest slopes, he has planted Maréchal Foch, a workhorse red that ripens early and produces good colour. "The vineyard challenge now is to find what can and cannot be done on that property," Watt says. "If it works here, there are many other places on the island where you can grow grapes. We have met several people here who are expressing interest in growing grapes. Ideally, it would be wonderful to have a Pender Island label."

On Mayne Island, just north of both Saturna and Pender, Alan Vichert and partner David Irwin

LIMITED PRODUCTION WINES ARE HAND-BOTTLED AT SATURNA ISLAND VINEYARDS.

REBECCA PAGE
SATURNA ISLAND VINEYARDS AND WINERY

acquired the 240-acre (97-hectare) Glen Echo Farm in 2001. The site of one of Mayne's former tomato greenhouses, it has a south-facing slope backing against a cliff that acts as a daytime heat sink, similar to the cliff on Saturna. The difference, Vichert says, is that his property is well inland, away from the cooling effect of the evening's sea breezes. While there is the potential for a vineyard larger even than Saturna's, Vichert cautiously prepared in 2002 to plant just five acres (two hectares). "At the beginning, we'll start with what seems to work here," he says. Initial plantings include Ortega, Pinot Gris, Auxerrois and Pinot Noir. Vichert also expects to plant some experimental varieties for the University of British Columbia's wine research centre.

The partners both have financial backgrounds. Irwin is a mortgage broker and Vichert has been a banker. Both also run an investment fund specializing in wine-related investments, notably in gene research. They had searched for several years for a vineyard property and were considering the Okanagan when they spotted a realtor's advertisement for the Mayne Island property in a wine publication. "This is a fabulous place for what we want to do," Vichert believes. For him, developing a vineyard and ultimately a winery is a chance to do something "completely different" in his life. The son of Baptist missionaries, he was born in China in 1942 and grew up in India and the Middle East, where he got an arts degree from Hebrew University in Jerusalem. Vichert went on to earn degrees in geology and business at the University of British Columbia. Through the 1990s, he was one of the two executives running the International Financial Centre in Vancouver. "All my life I've loved wine," Vichert says. "I was born with the taste. All those Baptists in my background never used their wine genes, so I got them all."

Named after a British frigate posted to the west coast in 1851, Thetis Island is much more remote in time than the 25-minute ferry ride from Chemainus would suggest. The baronial home at Overbury Farm has been in the Kasting family for four generations, since it was built in 1906. People who summer at the resort — one guest in 2001 had returned for 48 consecutive years — value the island's peacefulness and the night-time silence. Arlene and Norman Kasting's forebears exported poultry to Vancouver Island from their farm until, during the Depression, shipping services became irregular. That was when the family developed a resort with four rustic cabins, still equipped with wood stoves, kerosene lamps and period wooden outhouses. More modern accommodation has been added without destroying the experience of stepping back in time.

Norman Kasting, a neuropsychologist who taught at the University of British Columbia, and Arlene, a specialist in early childhood education,

have managed Overbury since 1990. Five years later, to maintain agriculture at the farm, they began three small vineyards, now totalling four acres (1.6 hectares), either close to the sea or within the forest. They planted Chardonnay, Pinot Blanc and Shiraz. These are challenging choices for a cool climate; the Shiraz does not ripen well every year. "We weren't very scientific in picking the varieties," Arlene says. "We chose what we like." The winery will be as low-key as the rest of the resort. "We don't have a vision to become a huge commercial operation. Our goal would be to have a market for the wines with our guests."

Colin Sparkes is more ambitious for his Thetis Island Vineyards. It is based on a six-acre (2.4-hectare) vineyard begun in November 2000 in a clearing well inland from the sea. The vineyard has a fine southern exposure, including terraces on which some late-ripening varieties have been placed. These include Cabernet Sauvignon (delivered in error by his supplier of vines). His other reds include Pinot Noir, Agria, Dornfelder, Shiraz and Pinotage. The whites are Gewürztraminer, Chardonnay, Sauvignon Blanc and Pinot Gris. Because Thetis — like most of the islands — is short of water, Sparkes has a 200,000-gallon (900,000-litre) irrigation reservoir at the top of the vineyard, not far from where space has been set aside for tennis courts. He intends to offer his guests amenities in addition to wine.

This mirrors his taste. Born in London in 1957, he was climbing mountains when he was 21 and has a peak in the Yukon named for him. After first studying and lecturing in agricultural engineering, Sparkes went on to study robotics. That led to a job

PREVIOUS PAGES: **SATURNA ISLAND'S REBECCA VINEYARD IS ONE OF THE LARGEST GULF ISLAND VINEYARDS.**

TRADITIONAL HARVEST FESTIVALS CELEBRATE THE VINTAGE.

in Heidelberg with SAP, the giant German industrial software company, where he met his wife, Carola Daffner, also a software consultant. She had grown up amid German wine regions and is even more passionate about wine than Sparkes. Their desire for the vineyard lifestyle was sealed by a two-year posting to San Francisco, which gave them many weekends in the Napa Valley and occasional trips to British Columbia. "We had the blood of the west coast in our veins," Sparkes says. They decided they wanted to develop what he calls an "adventure business" and found Thetis Island in 1999 after a three-year search for properties. Daffner has continued to work as a consultant while Sparkes has acquired the skills to run a resort and plant a vineyard. He dipped a toe into the art of fermentation by making apple cider in 2001.

In contrast to the other islands, Salt Spring is a well-serviced tourism community with good restaurants and perhaps 100 bed and breakfast establishments, many quite luxurious. Among its 9,279 residents (according to the 2001 census) is an unusual concentration of writers, artists, activists and other high achievers, setting high standards for each other as well as for the island's new wineries. "Salt Spring has developed a reputation of doing things nicely and properly," says Bill Harkley of Salt Spring Vineyards. Marcel Mercier adds: "What we are doing here on Salt Spring is very complementary to and synergistic with the other businesses here, such as the food businesses — cheese, lamb, pork — the bed and breakfasts, all of the agritourism." His Garry Oaks Estate Winery and Salt Spring Vineyards are neighbours on the Fulford-Ganges Road, the most heavily travelled road in the south end of Salt Spring.

Slightly off the beaten track, Long Harbour Vineyards has been developed on a farm owned by Bruce Smith, a real estate developer in his native Victoria where he was born in 1951, and his wife, Janet. The venture was launched after a 1994 vacation to Puligny-Montrachet in Burgundy. "We fell in love with the ambiance," he says. The farm is 175 acres (71 hectares) and extends down to Long Harbour. Formerly an orchard and then a sheep farm, it was acquired in 1961 by Janet Smith's American father primarily as a recreation property. The Smiths had taken it over just prior to their magical time in Burgundy. The year after that vacation, the first 500 vines were planted on the low-lying but still sloping property. A subsequent vacation to Tuscany cemented the vineyard commitment. When the eight acres (3.2 hectares) available for vines are fully planted, Long Harbour will be growing three clones of Pinot Noir, Maréchal Foch, St. Laurent (a red popular in Austria), Chardonnay, Pinot Blanc and Pinot Gris. Smith discarded an early choice, a red hybrid called Costel, because he disliked the herbal, rustic wine that it yielded. Since 1999, Paul Troop, arguably the most accomplished home winemaker on Salt Spring, has been making Long Harbour's wines.

"We're pushing the envelope here," Smith says of growing grapes at his site, one of the cooler vineyards on Salt Spring. Troop and Smith have confronted the issue head-on by experimenting with sparkling wines. "The specifications here are very good for bubbly," Smith maintains. He describes the Chardonnay table wine as "very steely." They have also produced rosé wine with the Pinot Noir. "Reds will always be a problem on the coast," he believes. "We'll be like every other small winery — bringing in Okanagan fruit."

At the Garry Oaks Estate Winery, Mercier and Kozak have been meticulous in their preparation. "The first thing I did — this is my systems approach of looking at problems — is that I got all the soil maps and climate maps and topographical maps and really assessed that quite carefully," Mercier says. Almost every single vine has been grafted onto roots specifically developed to accelerate the ripening of the grapes. "What we're striving for is to grow the best grapes we can for this particular climate and site and then make the best wine we can," he says. The varieties planted include Gewürztraminer, Pinot Gris, Pinot Noir and Léon Millot, a French hybrid grown because its deep colour is useful for darkening light reds. "Even in Burgundy, they used to refer to Léon Millot as a doctor of wine," Mercier notes.

They had owned property previously on two other Gulf Islands, but when they changed careers in 1999, they chose to plant a vineyard on well-populated Salt Spring. "We wanted to be in a healthy environment," Mercier explains. "We wanted a very strong intellectual challenge and we wanted a very strong business challenge." The seven-acre (2.8-hectare) vineyard was planted, beginning in 2000, on a gentle south-facing slope of a formerly heavily treed farm that was actually zoned as a gravel pit. Mercier speculates that the vineyard now covers a million dollars worth of gravel, except that a gravel pit would be inconceivable along Salt Spring's main road. "We'd be tarred and feathered and keelhauled," Kozak acknowledges.

Kozak, who grew up on a farm near the small Alberta town of Andrew, is an economist and was a vice-president with British Columbia Trade Development Corporation, a Crown agency dissolved in 1995. Mercier, with a postgraduate science degree, was a consultant in land and environmental management and has worked on projects around the world. "My aptitude test told me I should have been a farmer," he smiles. "I've always had my fingers in the soil." He recounts that, as a 10-year-old, he grew 50-pound (22.7-kilogram) prize-winning pumpkins in the family garden in his native Edmonton. Kozak and Mercier prepared thoroughly for the winery launch in 2003. "Once we got on this track, I figured out what I needed to know," says Kozak. She earned the higher certificate from the Wine and Spirit Education Trust and a diploma in viticulture and enology from the University of Guelph, with an eye to marketing the wines as well as working with a consulting winemaker at Garry Oaks. The ultimate intention is that the wines will be grown on the estate or from island grapes. However, the initial wines included a Pinot Noir made from Okanagan grapes. "We want wine that is a reflection of place," Kozak says. "The economic reality is that we can't afford to wait seven years until our vineyard is in full production."

At Salt Spring Vineyards, another 2003 launch, Bill and Jan Harkley prepared themselves by taking extension courses in vineyard and winery operations from the University of California at Davis. Born in Vancouver in 1940, Bill Harkley was a pilot and instructor until reaching the airline industry's mandatory retirement age in 2000. Jan, who was born in Calgary in 1953, is a chartered accountant and a business executive coach. When they bought their island farm in 1997, they decided they wanted to do something different there. "Everybody on Salt Spring does bed and breakfast," Jan observes (as, indeed, does this winery). Their three acres (1.2 hectares) of vines include Pinot Gris, Chardonnay, Pinot Noir and Léon Millot. Like Garry Oaks, they have also planted experimental early-ripening red varieties developed by a Swiss grape breeder. They have sought technical help on winemaking both from Paul Troop and from Chris Ottley, a Vancouver Island winemaker. It is entirely probable that this high-energy couple will acquire enough skills to handle most of their own winemaking. "We are not afraid to learn," Bill Harkley says.

GULF ISLANDS VINEYARDS

THE WINERIES

Garry Oaks Estate Winery
1880 Fulford-Ganges Road, Salt Spring Island,
BC, V8K 2A5
Telephone: 250 653-4687
www.saltspringwine.com

** Long Harbour Vineyards
301 Mansell Road, Salt Spring Island, BC, V8K 1P9
Telephone: 250 537-2904
E-Mail: dbrucesmith@shaw.ca

Morning Bay Farm
6621 Harbour Hill Road, North Pender Island,
BC, V0N 2M1
Telephone: 250 629-8350
E-Mail: keith_watt@mybc.com

** Overbury Farm Resort
288 Forbes Road, Thetis Island, BC, V0R 2Y0
Telephone: 250 246-9769
www.overbury.bc.ca

Salt Spring Vineyards
151 Lee Road, Salt Spring Island, BC, V8K 2A5
Telephone: 250 653-9463
www.saltspringvineyards.com

Saturna Island Vineyards and Winery
8 Quarry Road, Saturna Island, BC, V0N 2Y0
Telephone: 250 539-5139
www.saturnavineyards.com

Thetis Island Vineyards
90 Pilkey Point Road, Thetis Island, BC, V0R 2Y0
Telephone: 250 246-2258
www.cedar-beach.com

** UNDER DEVELOPMENT

FRASER VALLEY VINEYARDS

"WE'VE WALKED AROUND VINEYARDS THAT USE HEAVY HERBICIDES AND PESTICIDES AND WALKED AROUND OUR OWN — AND OUR OWN IS FULL OF BUTTERFLIES AND BEES AND LADYBUGS."

—DAVID AVERY
A'VERY FINE WINERY

The winery that established the Fraser Valley as a credible wine-growing region was Domaine de Chaberton. It opened in 1991 south of Langley, almost at the international border. Based on the valley's largest vineyard, it was for almost a decade the only winery in the valley that processed grapes or fruit grown there. Columbia Valley Classics, British Columbia's first fruit winery, opened at Cultus Lake in 1998. (Andrés winery has operated in Port Moody, a Vancouver suburb, since 1961 but never had a vineyard in the Fraser Valley.) Domaine de Chaberton got company in the summer of 2001 when Township 7 Vineyards and Winery opened just around the corner and The Fort Wine Company, a second fruit winery, arose from a cranberry farm east of Fort Langley. The cluster of wineries increased again in the summer of 2002 when Glenugie Winery and A'Very Fine Winery opened, cheered on by the owners of the dozen other modest valley vineyards, some of which also nurse winery ambitions. "We are happy to get people here who open wineries as long as they make good wine," says Ingeborg Violet, co-proprietor with husband, Claude, of Domaine de Chaberton. "The Fraser Valley has a good reputation."

The fertile delta south of the Fraser River is Vancouver's agricultural hinterland, where extensive dairy, livestock and berry farms flourish, along with occasional specialty crops (one historic farm

OPPOSITE: **TIGHTLY STACKED BOTTLES OF CLASSICALLY MADE SPARKLING WINES, A SPECIALTY OF MANY BRITISH COLUMBIA WINERIES.**

produced hops for beer for almost a century). By 1979, when Claude Violet first surveyed possible vineyard sites, the sprawl of Vancouver's growth was consuming property and driving up land prices. Vineyard land was cheaper in the Okanagan; few aspiring wine growers gave the Fraser Valley a second thought. Estate wineries were impractical then because each was required to have its own 20 acres (eight hectares) of vineyard — an expensive proposition in the Fraser Valley. The current land-based winery rules, requiring fewer acres, enable new wineries to open, even though land prices remain high. The 10-acre (four-hectare) A'Very property just west of the Abbotsford airport adjoins a 100-acre (40-hectare) hay meadow on a reclaimed gravel pit. One of the single largest unexploited farms in the valley, it might be suitable as a vineyard — except for the $5 million asking price.

Claude and Inge Violet of Domaine de Chaberton brought money and experience to wine growing. In 1866, an ancestor, Simon Violet, who is dismissed in one reference book on spirits as a "shepherd of the Pyrenees," created an apéritif wine called Byrrh, a bitter-sweet digestive. Far from being just shepherds, the Violet family had been growing grapes and making wine in southwestern France since the middle of the seventeenth century. Byrrh became so enormously popular that production reached 100,000 bottles a day in 1935, some of it aged in the world's biggest wooden cask (one million litres). Claude Violet's father, Jacques, merged his company with those of two other apéritif producers in a transaction that left the family considerably better off than their shepherding days. Claude, who was born in Paris in 1935, met Inge while he worked at a bank in Munich. "It is a good thing to know a little bit about how a bank is working," he once muttered sardonically after Domaine de Chaberton had been open a few years. In the 1960s, he ran a large vineyard near Montpellier. (Domaine de Chaberton also was the name of the family's farm in France.) He and Inge then established a wine import business in Switzerland until, wary of Europe's Cold War tensions, they decided to move to North America and make wine.

Their thorough search spanned American wine regions from New York to California but they preferred Canada. "I liked the U.S. and I liked the Americans," Claude told a journalist in 1990. "But if you did not know anything about baseball or what the local university football team was doing, you were lost. The U.S. is a nice place to visit but I didn't want to live there." They ruled out vineyard property in the Niagara because, in Claude's judgment, Ontario winters were too cold for vinifera grapes while the hybrid and labrusca grapes then common there made poor wine. They looked at British Columbia's Okanagan, judging it to be too far from Vancouver's big wine market. Claude also thought the Okanagan winters might be too harsh; the unusually cold winter of 1978–79 had severely damaged vineyards in both the south Okanagan and the adjacent Similkameen Valley. In the Fraser Valley, he met with John Harper who, with a trial vineyard near Cloverdale, had become an authority on growing grapes. Violet learned that the valley, with the Vancouver market on its doorstep, has mild winters and long summers, with heat units comparable to Colmar in the Alsace wine-growing region.

In the summer of 1980, Violet looked for vineyard sites from White Rock to Cultus Lake — "not every avenue and every street, but a lot of them," he recalls. He checked regularly with Harper; finally the veteran grape grower approved of a farm on 216th Street just north of the 49th parallel. From weather data, Violet knew that the farm was in one of the Fraser Valley's dry pockets. The underutilized 55-acre (22.3-hectare) farm grew only small quantities of strawberries and raspberries. Violet bulldozed a gentle south-facing slope for his 40-acre (16-hectare) vineyard. "The slope was already there," Claude says. "We just redid it into a perfect slope."

Planting began in 1982 with Bacchus vines imported from a nursery in Germany. The variety had been bred there a decade earlier to mature in cool climates. "I saw Bacchus growing in Germany in places where the climate is not very good," Claude says. "So I concluded it would grow here. I was right. Our Bacchus is doing very well." (The

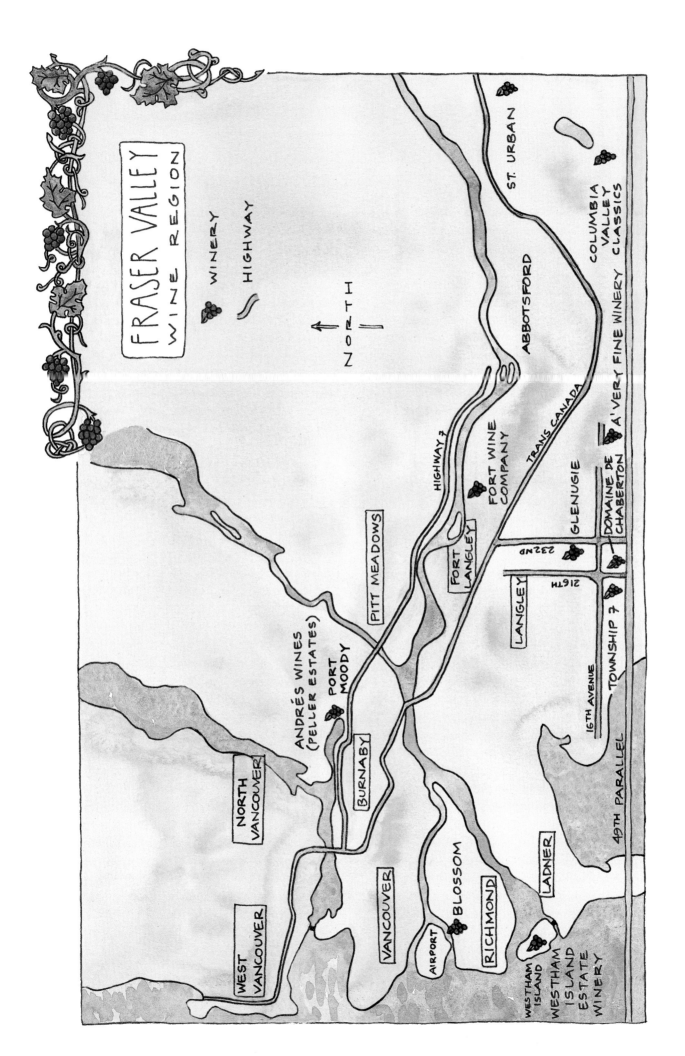

variety also yields light, crisp, aromatic wines on Vancouver Island and near Shuswap Lake.)

The Fraser delta's climate is not difficult, provided that wine growers are as careful as Violet to choose suitable varieties. Many white varieties and some early-ripening reds, including Pinot Noir, are grown successfully. The sun is seldom intense enough to produce full-bodied Cabernets or Merlots (though these are planted occasionally). Several vineyards arrange plastic sheets around some vines each spring, creating a temporary greenhouse to accelerate vine growth by two or three weeks. However, all Fraser Valley wineries also buy grapes from Okanagan vineyards; the Violets even financed a grower to plant a vineyard on Black Sage Road near Oliver.

In addition to Bacchus, Claude Violet's initial plantings also include Madeleine Angevine and Madeleine Sylvaner, white varieties developed in the Loire in the nineteenth century and now common in the cool vineyards of England. At Domaine de Chaberton, these varieties, along with Bacchus, Optima and Ortega, also early-maturing whites, have become signature wines since the winery opened in 1991. By its tenth anniversary, Domaine de Chaberton was an established medium-sized winery, making about 25,000 cases a year and selling half of it directly through the well-patronized wine shop and in the winery's bistro.

Domaine de Chaberton's success attracted Corey and Gwen Coleman when they planned Township 7. Both were working at Okanagan wineries but concur with Violet's reason for locating in the Fraser Valley. "The real key was to be closer to our customers, the lower mainland being the main market for B.C. wines," Corey Coleman adds. "We felt it was easier to move the grapes here than it was to try to get the people out to the Okanagan. We could also have more direct contact with our customers. We wouldn't need to rely on agents as much and we could have that personal contact."

While Claude Violet represents the ninth generation of his family growing wine, the Colemans are newcomers. Born in Saskatchewan (Corey in 1964, Gwen in 1965), they met while getting marketing degrees at the University of Saskatchewan. A 1989 vacation to California's Napa Valley lit a passion for wine so profound that subsequent vacations were all taken in wine regions. During 18 months in Montreal when Gwen worked for a drug company, they even toured the small wineries in Québec's eastern townships. When they returned to western Canada, Corey Coleman sent his resume to 23 wineries. In his first job at the Tinhorn Creek winery near Oliver, he was part of the crew that planted the winery's Black Sage Road vineyard. He began learning winemaking skills from Tinhorn's Sandra Oldfield before moving in 1997 to Hawthorne Mountain Vineyards where he had a hand in three vintages as well as making Township 7's debut wines. These included some 312 cases of 1999 Merlot with Okanagan grapes and 4,000 bottles of a sparkling wine, made opportunistically when Coleman found some Pinot Noir and Chardonnay grapes going unused at another Okanagan winery.

Meanwhile, Gwen Coleman honed her wine marketing skills by working with Sumac Ridge founder Harry McWatters. "I don't know who else you could think of to be a better tutor than Harry as far as marketing and sales go," Corey believes. "He's the king." Not having the equivalent of the Byrrh fortune behind them, the Colemans started small in November 1999, purchasing a five-acre (two-hectare) parcel, including a now-renovated 1929 farmhouse, on 16th Avenue, one of the longest and busiest through streets south of the Fraser River. The winery name was inspired by a legal description on an 1886 property map referring to the region as Township 7. "It was put on a list with five other names that we shopped around to our friends and associates," Corey recounts. "Then our business side took over. It's easy to say. It's easy to remember. It's distinctive and easy to spell — all those marketing issues that are important." The Colemans are fond of the number seven (they opened their winery on the seventh day of the seventh month in 2001, and they call their sparkling wine Seven Stars).

The Colemans have their life savings on the line. Township 7's desirable location comes at a cost.

Corey calculates that, for the same purchase price, he might have acquired four times as much land in the Okanagan. About three-quarters of the parcel has been planted to vines since the spring of 2001; the varieties are Pinot Noir, Chardonnay, Optima and "some Merlot just for fun." Coleman has contracted Merlot, the winery's signature red, and other grapes including Syrah, Pinot Gris and Sauvignon Blanc, from growers in the south Okanagan. "We're very up front with people," he says of his use of Okanagan grapes. "Our goal is to make the best wine from the best British Columbia fruit we can get." By the 2001 vintage, Coleman was making 2,000 cases of wine, with the intention of doubling in size as the market discovers his full-flavoured and well-made wines.

The additional Fraser Valley wineries that opened in 2002 benefit both Domaine de Chaberton and Township 7 by creating a cluster to attract wine tourists. The Glenugie Winery (pronounced glen-yew-gie) is only a 17-minute drive east and north from Township 7 while the new A'Very winery is 45 minutes straight east on 16th Avenue (which is renamed King Road when it crosses the municipal border into Abbottsford). The organic five-acre (two-hectare) vineyard behind Glenugie is planted exclusively to Pinot Noir, a particular favourite of Gary Tayler, who established the winery with his family. Born in Edmonton in 1939 and the son of a printer, Tayler planted a vineyard south of Penticton in 1976 in which he grew primarily Riesling but also some Pinot Noir. When the depressed grape prices of 1988 made the vineyard uneconomic, Tayler moved to Langley, returning to his previous occupation as a contractor and developer. He is now retired from that business but his prowess shows in the quality workmanship of

UNRULY VINES AT DOMAINE DE CHABERTON ARE PRUNED TO THE BUD WOOD NEEDED FOR NEXT YEAR'S CROP.

COREY AND GWEN COLEMAN
TOWNSHIP 7 VINEYARDS AND WINERY

CLAUDE VIOLET
DOMAINE DE CHABERTON

GARY TAYLER
GLENUGIE WINERY

KIRK SEGGIE AND PHILIP SOO
ANDRÉS WINES

the handsome ochre-toned, brick-clad steel and concrete winery with a capacity to produce 10,000 cases a year. He calls the building "a Scottish fortress," in part because of the unusual slit windows on both the ground and second floors. Slightly narrower than a head, the windows frustrate break-ins.

In 1997, when Tayler began planting his Langley property, he intended to put in just enough vines to make Pinot Noir for his family. "The 25 vines exploded to 8,700," he recalls, and he could not find a market for his grapes. The family was polled and the Taylers decided to open a winery. One daughter, Michelle, who has a degree in agriculture, looks after the winemaking while the other daughter, Lara, a commerce graduate who specialized in marketing, looks after wine sales. The books are done by their Scots mother, Christine, whose heritage is honoured in the winery name. Her forebears had a farm in a glen in eastern Scotland near the Ugie River. (There was also an unrelated Glenugie single malt distillery which closed in 1982.) The winery processes Tayler's Pinot Noir and also buys Okanagan grapes, including Chardonnay and Merlot. The winery's Limited Vintage Pinot Noir 2001, dark and full-bodied, comes from a vineyard in Osoyoos. Tayler is so pleased with the wine that he will not disclose the grower's name. "I don't want anybody else chasing him," he says.

A'Very Fine Winery's name coyly plays on the surname of the owners: David and Liesbeth (Liz) Avery, both keen amateur winemakers. David was born in Toronto in 1955 and, among other occupations, was general manager of an office equipment supply company for 13 years. Liz Avery was born in Paraguay, the daughter of a farmer who immigrated to Canada in 1973 with his family. The Avery property just west of Abbottsford, which they acquired in 1996 as a hay meadow, is a former gravel pit that was reclaimed with a laser-graded slope, four percent due south and ideal for a vineyard. They began planting in the spring of 1998. It was an unusually hot, dry summer and, having been advised that irrigation is unnecessary in the Fraser Valley, they scrambled to water the tender new vines. "A sprinkler that covers your front lawn just looks like a toy in an eight-acre vineyard," David says. They installed drip irrigation and brought most of the vines through that dry season.

About half the vineyard is planted to Pinot Noir; there are also smaller plantings of Pinot Meunier and Chardonnay. "If we ever did want to get into sparkling wine, we have the ingredients," David says. The vineyard also includes Gewürztraminer and Siegerrebe, two of David's personal favourites, as well as Pinot Gris, Gamay and, on a steep south-facing bank below the winery, Cabernet Franc. With the ambition of becoming a 10,000 case winery, the Averys also buy Okanagan grapes.

Both Glenugie and A'Very offer, among other wines, unoaked Chardonnays remarkably similar in style. That is not coincidental: both wineries have the same consultant, Elias Phiniotis, who has also backed up Claude Violet at Domaine de Chaberton since 1991. Born in Cyprus in 1943, Phiniotis earned a doctorate in food chemistry in Hungary before coming to Canada in 1976. Based in Kelowna, he has made wine for a number of wineries, including Calona and Quails' Gate, and has consulted as far afield as China and Sri Lanka.

Paul Kompauer is going it alone at St. Urban Vineyards at Chilliwack. "I am a seventh-generation winemaker from Slovakia," he says. "I made my first wine when I was 12 and it was drinkable." Kompauer was a 19-year-old student in the former Czechoslovakia when the Soviet troops invaded in August 1968 to snuff a burgeoning democracy. He fled to Canada, resumed his studies and graduated as a civil engineer in 1976. Ten years later, he established his own firm, Inter-Coast Consultants, specializing in building restoration. Throughout a successful business career, he has wanted to resume the family tradition of growing wine. An offer he made on a major Similkameen Valley vineyard slipped away because he could not raise the cash after failing to sell his Surrey home. Finally, in 2001 he purchased a Chilliwack property with seven acres (2.8 hectares) of reasonably mature vines.

The previous owner had planted the vineyard with a winery in mind and had started making wine. Kompauer, who took over after the project collapsed, discovered a dozen cases of Agria had been left behind in the crawl space of the house. The deep, full-bodied red wine was promising. "It needs a little oak," he says. "I like oak in my wines." The other four varieties in the vineyard all are early-maturing whites appropriate to Fraser Valley viticulture. The name of the winery, St. Urban, might prove a talisman against spring frosts. One day late in May several years ago in Slovakia, Kompauer happened on a celebration at a wine museum marking the feast day of St. Urban. He learned that this is the patron saint of vineyards and that, by tradition, there is no frost after May 25, the saint's day in central Europe. It was, he decided, the ideal name for his winery, which he expects to open in mid-2003.

The Fraser Valley's abundant fruit and berry production has given rise to several wineries since the 1998 opening of British Columbia's first modern-era fruit winery, Columbia Valley Classics. It is a 45-acre (18.2-hectare) farm and orchard on a hillside overlooking Cultus Lake, the popular resort about 90 minutes east of Vancouver. The winery's tasting room and retail shop offers wines from blueberries, red, white and black currants, raspberries, gooseberries, saskatoon berries, rhubarb and kiwi fruit. For good measure, there are conserves made from estate-grown hazelnuts along with a hazelnut liqueur, introduced in 2002. The winery also produces some wines from purchased grapes.

Columbia Classics was conceived by John Stuyt, a rugged farmer born in Holland in 1930 to a family

YOUNG VINES FROM NURSERY READY FOR PLANTING AT THE TOWNSHIP 7 VINEYARD.

with a history of at least 350 years in agriculture. After immigrating to Canada in 1956, Stuyt grew everything from raspberries to chickens at a farm near Aldergrove. One Sunday in 1989, while out for a casual drive, he came across the Columbia Valley and was so impressed with its potential for agriculture that he bought what was basically raw land within five days. He began planting it the next year. "This little valley will grow everything better than anywhere else," he maintains. It is a rich, sun-drenched terrain surrounded by low mountains that radiate the day's heat across the fields in the evenings. The lake moderates the climate early in the season when frost might be a risk. Once his berry bushes and nut trees began producing, low prices led Stuyt to look for ways to add value to his crop.

His was the first fruit winery to open in British Columbia in almost 80 years. "The fruit wine industry is just beginning to be recognized but we are here to stay," Stuyt says. In 2001, the Fraser Valley's second fruit winery opened, strategically located a few minutes east of Fort Langley. The fort, the re-creation of a nineteenth century Hudson's Bay Company trading post, is a federally operated heritage attraction drawing thousands of visitors each year. Some spill over to taste the cranberry wine offered by The Fort Wine Company, whose newly built winery echoes the architecture of the trading post. The theme is also captured in the name of the winery's fortified raspberry wine — Raspberry Portage.

This winery is the creation of Wade Bauck and Terry Bieker. Bauck, born in 1960, captains a towboat in the Port of Vancouver. In his spare time, he has developed a five-acre (two-hectare) cranberry bog on his farm in the Glen Valley east of Fort Langley, selling his crop profitably each year to the leading cranberry processor. However, a slump in berry prices late in the 1990s led Bauck, like Stuyt, to look at other ways to add value. "Wine seemed the most viable option," he concluded. He enlisted the participation of a friend, oil refinery manager Terry Bieker, in the winery, which opened in the summer of 2001. Neither partner had much experience in the wine business; Bieker, seven years older than his partner, quips that he is an avid consumer.

GRAFTED GRAPE PLANT BURSTS INTO LIFE.

They took a crash course by visiting the Napa Valley before they opened their winery.

To make their wines, they hired Dominic Rivard, one of British Columbia's most experienced fruit winemakers. Born in Québec's Gaspé region in 1971, Rivard began making wine in his teens and honed his skills working in the laboratory of Spagnol's Enterprises, a major Vancouver supplier of grapes and wine supplies for amateurs. Rivard was the initial winemaker at Columbia Valley Classics and briefly made fruit wine in the Okanagan before joining The Fort. While this winery has also made wines with purchased grapes, it has concentrated on fruit wines, notably cranberries. "It's not an easy wine to make," Rivard says. The tiny round berries, with little juice and hard skins, must be crushed in a hammer mill before fermentation. The acidic juice is balanced with the addition of water and sugar. Rivard produces a dry base wine with a maximum 11 percent alcohol and then balances the wine with a touch of sweetness. The result is piquant, zesty and remarkably like grape wine. He has also added a line of fortified fruit wines, an iced apple wine and specialty products ranging from conserves to cranberry balsamic vinegar.

The Blossom Winery, which opened in December 2001 in downtown Richmond, is located somewhat incongruously in a strip mall with a furniture store as a neighbour. However, the attractive tasting room more than compensates for the commercial exterior. The winery is the dream of Taiwanese-born John Chang and Allison Lu, who operated a wholesale electrical supply business in Taipei before settling in Canada in 1999. A self-taught winemaker who still cherishes his grandmother's recipes for fruit wines, Chang was impressed with the taste of Fraser Valley blueberries and raspberries. Soon after arriving in Canada, he started to produce trial lots of wine and, when satisfied with the outcome, proceeded to develop the winery. "Some people think that fruit wine is not high quality," says Allison Lu, translating for her husband. "John's desire is to have high-quality wines." The wines display attractive aromas and flavours of fresh fruit. Both the Blueberry Reserve and the Raspberry Reserve are just slightly sweet food wines, while the Blueberry Late Harvest and the Raspberry Late Harvest are finished as dessert wines.

The Westham Island fruit winery was conceived by Andy Bissett, a friend of John Stuyt at Columbia Valley Classics winery. Bissett was a long-time fruit grower in the Fraser Valley with farms on Westham Island (at the mouth of the south arm of the Fraser River) and near Agassiz. The winery plans were well advanced, with most of the approvals in hand from the municipality of Delta, when Bissett died suddenly early in 2002. Lorraine Bissett, his widow, has taken her husband's dream to completion, converting a barn on the family's 25-acre (10-hectare) Westham Island farm into a winery and tasting room and recruiting Ron Taylor as the winemaker. A microbiologist who spent 22 years as the winemaker at Andrés in Port Moody before becoming an independent consultant, Taylor previously helped make the fruit wines at Blossom. The winery plans to produce wines from blueberries, raspberries, currants, boysenberries, blackberries, marionberries, rhubarb and even tayberries (a 1977 blackberry-raspberry cross).

This winery is conveniently located on the road leading to the renowned George C. Reifel Migratory Bird Sanctuary on Westham Island. Reifel was a Vancouver brewer and real estate developer who began buying farm property on the fertile island in the 1920s. A bird sanctuary was established there in 1961 because the island is on a major flyway for migratory birds. Subsequently, the Reifel family gave a portion of its Westham Island holdings to the federal government to expand the sanctuary, now a primary year-round location for bird-watching in the Vancouver area. Lorraine Bissett's wine shop, which also sells jams, jellies and syrups, is sure to attract birders — particularly as a warm stop after watching the snow geese in November.

FRASER VALLEY VINEYARDS

THE WINERIES

Andrés Wines (Peller Estates)
2120 Vintner Street, Port Moody, BC, V3H 1W8
Telephone: 604 937-3411
www.andreswines.com

Blossom Winery
5491 Minoru Boulevard, Richmond, BC, V6X 2B1
Telephone: 604 232-9839
www.blossomwinery.com

Columbia Valley Classics
1385 Frost Road, Lindell Beach, BC, V2R 4X8
Telephone: 604 858-5233
www.cvcwines.com

Domaine de Chaberton
1064 216th Street, Langley, BC, V2Z 1R3
Telephone: 604 530-1736
www.domainedechaberton.com

The Fort Wine Company
26151 84th Avenue, Fort Langley, BC, V1M 3M6
Telephone: 604 857-1101
www.thefortwineco.com

Glenugie Winery
3033 232nd Street, Langley, BC, V27 3A8
Telephone: 604 539-9463
www.glenugiewinery.com

Lotusland Vineyards
(formerly A'Very Fine Winery)
28450 King Road, Abbotsford, BC, V4X 1B1
Telephone: 604 857-4188
www.averyfinewine.ca

**St. Urban Vineyards
47189 Bailey Road, Chilliwack, BC, V2R 4S8
Telephone: 604 858-7652

Township 7 Vineyards and Winery
21152 16th Avenue, Langley, BC, V2Z 1K3
Telephone: 604 532-1766
www.township7.com

Westham Island Estate Winery
2170 Westham Island Road, Delta, BC, V4K 3N2
Telephone: 604 946-7139
E-Mail: bissett@netcom.ca

** UNDER DEVELOPMENT

WINEMAKERS OFF THE BEATEN PATH

"IF YOU GET LUCKY, YOU FIND A SMALL DEDICATED RESTAURANT THAT REALLY, TRULY LIKES YOUR WINES AND THEY SELL IT BY THE CASE. THERE AREN'T TOO MANY OUT THERE LIKE THAT."
—HANS NEVRKLA
LARCH HILLS

Michael and Susan Smith opened the doors at Recline Ridge winery near Salmon Arm on the morning of July 16, 1999. Within an hour, the tasting room was closed by the local health inspector, who also threatened to confiscate all the wines. Believing that wine is made with the addition of water, he closed the winery because it received untreated water from a local creek. The Smiths reopened late that afternoon after an urgent appeal to the regulators in Victoria. They put bottled water into their tasting room and they agreed to spend $5,000 on a chlorinating system. "It was quite stressful for an opening day," Michael Smith recalls.

Contrary to what the inspector thought, Smith does not ameliorate his wines with water, which is forbidden in any event under the Vintners Quality Alliance standards. (Home winemakers, who far outnumber the professionals, typically add water to grape concentrates.) Salmon Arm (population 15,000), 81 miles (130 kilometres) north of Kelowna and the most northerly wine-growing region in British Columbia — perhaps in North America — is off the mainstream of wine. Dealing with ill-informed inspectors is the least of the challenges confronting isolated wineries like Recline Ridge. At times unheralded even in their own communities, these wineries strive in splendid loneliness, far from knowledgeable markets, in vineyards from

OPPOSITE: **WINE GRAPES ARE PICKED INTO SMALL BOXES AT THE LARCH HILLS WINERY.**

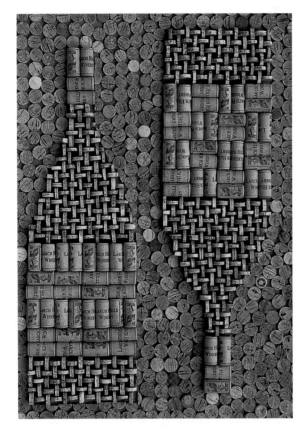

SEVERAL WINERY GIFT SHOPS SELL CORK ART CREATED BY NANCY SCOTT OF CHASE, BC.

Before Recline Ridge, the northernmost British Columbia winery was Larch Hills, perched high on a steep, south-facing vineyard midway between Salmon Arm and Enderby, just beyond the northern end of the Okanagan Valley. It still boasts the highest elevation (about 2,300 feet/700 meters) of any British Columbia winery. The winery, which takes its name from the surrounding forested hills, was opened in 1997 by Hans and Hazel Nevrkla. Both Michael Smith and Hans Nevrkla came to professional winemaking after many years as award-winning amateurs. In his laboratory, Smith still keeps the home winemaker's bible, a battered copy of The Art of Making Wine by Stanley Anderson, which was published in 1968 and has sold more than 350,000 copies.

Hans Nevrkla feels the isolation as much as the Smiths do. "We don't even get to hear all the gossip that goes on down there," he says, gesturing south toward the Okanagan. Born in 1946 in Vienna, a city surrounded by wineries, Nevrkla, an electronics technician, and his British-born wife, Hazel, came to Canada in 1970, settling in Calgary. Having grown up with palatable wine, he was unhappy with the quality of Canadian wine at that time. "Up until that point, I had never come across bad wine," he said. "I started making my own out of necessity." He became so skilled that he taught both winemaking and wine appreciation in Calgary's school system for more than a decade. He became familiar with the Okanagan during years of buying grapes from vineyards near Oliver. But when he and Hazel left Calgary in 1987, they purchased a mountainside property in the verdant Salmon Arm area because they thought the south Okanagan was "too brown." When a few vines planted near the house survived, Nevrkla decided to plant a vineyard. "Everybody said it wasn't going to work," he recalls. It seems that the community had forgotten that grapes were grown before in the region. In 1907 a Salmon Arm farmer named W.J. Wilcox planted a small vineyard of table grapes. And what is now the Raven subdivision in Salmon Arm was formerly a vineyard owned by a family called Raven.

the Pemberton Valley north of Whistler to the Columbia Valley near Trail. As the Smiths have discovered, it is difficult to attend industry meetings, to get capable vineyard help, to buy winery supplies or even to taste and compare wines with their peers, let alone attract consumers to out-of-the-way wineries. Because of their marginal locations, these wineries cannot grow fashionable varieties such as Merlot or Syrah (although most buy varieties not grown in their vineyards). Smith has found the VQA tasting panel unenthusiastic about wines from the obscure varieties often grown in isolated vineyards. "You'll send something into VQA," he says "and they'll come back with 'Not bad for a Chancellor' or 'Good effort for a Foch.' It's so obvious that if you are not moving with the larger winery trends where you are forced to make wines that are market driven, there is a stigma attached to it."

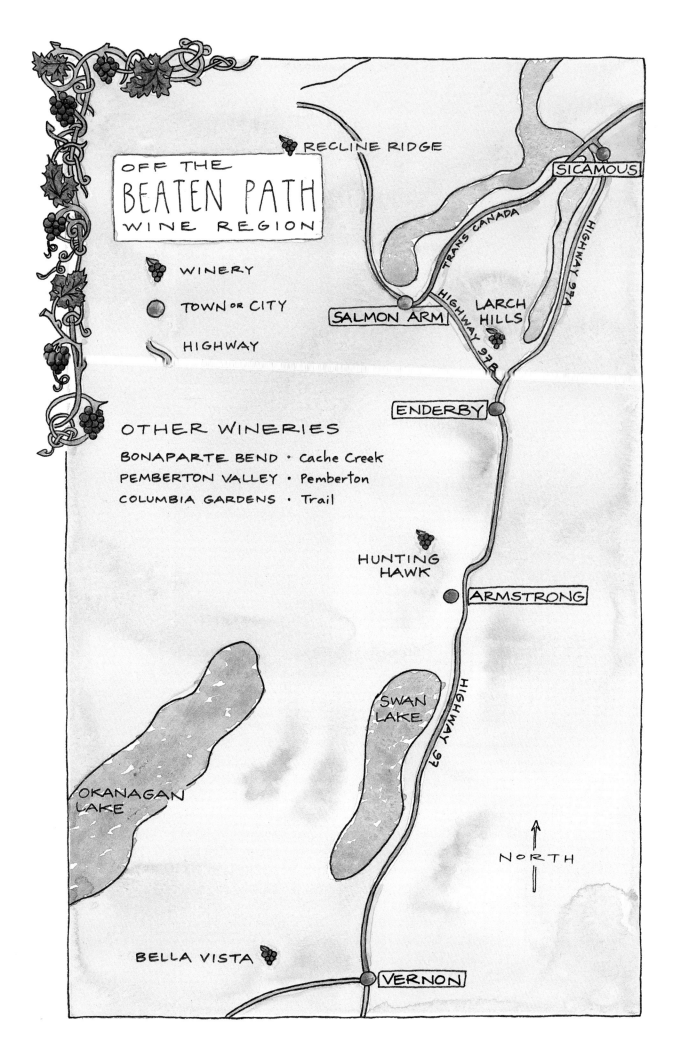

HAZEL AND HANS NEVRKLA
LARCH HILLS WINERY

SUSAN AND MICHAEL SMITH
RECLINE RIDGE VINEYARDS & WINERY

LARRY PASSMORE
BELLA VISTA VINEYARDS

Nevrkla started in 1989 with a quarter-acre (1000-square-metre) trial. Ortega, which proved the grape most suitable to his vineyard, now comprises about half of the six and a half acres (2.6 hectares) that he grows. He also has Madeleine Angevine and Siegerrebe, which produce singularly fruity and aromatic white wines. For what will be the winery's premier red wine, he is expanding a small planting of Agria, a dark, full-bodied Hungarian variety also grown on Vancouver Island. "This site is unique and we are growing the varieties that work here," he says. The vineyard, with its 15 percent slope to the south, has good sun exposure. The clay-loam soils retain moisture so well that, unlike most Okanagan vineyards, Larch Hills does not irrigate. However, the climate is challenging. In the fine warm season of 1995, Nevrkla picked perfect Ortega grapes on September 20. But in 1996, a notoriously cool year, he picked a month later and the grapes were ripe enough only for a light wine. Then heavy snows buried the vines before the leaves had fallen, damaging both the vines and the following spring's buds. Nevrkla also purchases grapes for the winery, which opened in 1997 and now makes about 2,500 cases a year. He buys Pinot Noir, Gewürztraminer and Merlot from growers near Westbank.

WINEMAKERS POUR THEIR WINES AT FESTIVALS THROUGHOUT THE YEAR.

"Why struggle with varieties that can be grown in more suitable spots?" he says. "I'm better off to get contracts with successful growers down south."

The Recline Ridge winery evolved from a rural property about eight and three quarter miles (14 kilometres) west of Salmon Arm. The Smiths, who both work in town (he runs the cable company, she is a planning department clerk), intended to pasture horses after clearing the land. It took them five years to get rid of the grazing grasses they had planted when they started developing the five-acre (two-hectare) vineyard in 1994. Born in Ottawa in 1949, Michael Smith was already a keen amateur winemaker before establishing Recline Ridge. He planted the same varieties that Larch Hills grows, plus Maréchal Foch. In 1998 he secured cuttings of what he was told were Gewürztraminer vines and was delighted when they thrived even after being planted during a drought. The vines were into their second year when a visiting consultant identified them as Concord. Disappointed, he pulled them out and replaced them with Siegerrebe, one of Recline Ridge's most successful wines. He backs up the production of his own vineyard with grapes from more southern Okanagan vineyards, including Chardonnay, Chancellor, Pinot Noir and Perle of Csaba. Smith has an eye on plantable property near Recline Ridge, should it come onto the market. Meanwhile, he can also buy grapes from local growers like Jim Wright. A retired teacher, Wright operates the three-acre (1.2-hectare) Ashby Point Vineyard on a property overlooking Shuswap Lake that has been in his family since 1906. His vines are on a warm, south-sloping bench that, due to its exposure and the lake effect, matures the fruit

MERLOT GRAPES PRODUCE SOME OF BRITISH COLUMBIA'S MOST POPULAR RED WINES.

earlier than the Recline Ridge vineyard. Wright is growing Ortega, Madeleine Sylvaner, Schönburger and Maréchal Foch and sells the grapes to both Recline Ridge and Larch Hills. Wright has additional land available for planting.

Smith is optimistic that one day he will be able to produce enough wines from the region's grapes to support a Shuswap sub-appellation. By 2001, Recline Ridge was making a total of 2,000 cases of wine. That is more than enough for the industrious Smiths, so fully occupied with their regular employment that tasks such as bottling wines often are done on quiet Sunday afternoons. "Somehow we have to get out of this cycle of perpetual work," Michael Smith says wistfully. "We never get the opportunity to spend any time with other people that are making wine."

Smith did, however, influence Barry Tunzelmann's decision to plant a four-acre (1.6-hectare) vineyard in 2001 near Canoe, B.C., just east of Salmon Arm. Tunzelmann, who grew up on a New Zealand farm and came to Canada in 1975 as a young man, is an electrical engineering technologist who has worked both in the British Columbia forest industry and in Alberta's oil and gas industry. In 1994 he and his family moved to an acreage at Canoe and began searching for ways to make it pay. "We had tried the animal route, pigs, beef, sheep and all that stuff, and we could barely get our money back," he says. "There was no way we could ever make a living off this acreage. You'd still have to supplement your living with permanent work." Then Michael Smith told him there was a demand for grapes grown in northern vineyards. Tunzelmann did his mathematics and concluded that he could make his acreage pay eventually by planning on a future winery. In 2001 he planted Siegerrebe, Ortega, Optima and Maréchal Foch. A further three acres (1.2 hectares) is reserved for more Foch and for Agria. Tunzelmann plans to call the winery Lyman Estates. He will retain a consultant to make the wine. "I want to hone my viticulture skills first," he says. "All my homework tells me that if you haven't got good grapes, you can't make good wine."

Sunday afternoons in the spring are likely to find Russ and Marnie Niles selling bedding plants from their shop in a supermarket parking lot in Armstrong. "If you don't have lots of money and want to get into the wine business, you have to be prepared for 14- to 16-hour days of hard slugging," he says. "Man, you work your guts out at this!" The cash from the plant business buys the bottles needed for their Hunting Hawk Vineyards. They began selling wines in late 2001 and opened a tasting and sales room (in their home) the following June. "I love wine and I love making wine," Russ Niles says, "but I also view this as a business opportunity. This is not my retirement dream."

Niles was born in Victoria in 1957 into an often-transferred air force family. He attended 14 different schools and when he completed his education, he wanted nothing more than to settle down. Soon after going to work at the *Vernon Daily News*, he and his wife bought rural property north of

Armstrong. "Here we built our house and raised our family on the salary of small-town journalism," he says. He was editor of the paper when its owners closed it. For another five months, he edited a new daily newspaper until it ran out of money. While continuing to work as a freelance writer, Niles, who had been a partner for a time in another winery, plunged into a winery of his own. Gulch Road, the country road on which Hunting Hawk is established, is aptly named. The three-acre (1.2-hectare) vineyard is planted on a terraced sand ridge that was heavily forested until Niles logged it in 2000. The sun-bathed southwestern slope descends steeply into a deep ravine. Beside Deep Creek, which flows through the ravine, is the Niles home and temporary winery until a dedicated winery can be erected at the top of the vineyard. Niles has planted Maréchal Foch, Ortega and Perle of Csaba, varieties which ripen readily in northern vineyards. Hunting Hawk also leases small vineyard parcels and signs up growers farther south in the Okanagan, enabling Niles to make such wines as Merlot, Chardonnay and Gewürztraminer.

Bella Vista Vineyards of Vernon is not particularly remote (it is perhaps five minutes off the highway from downtown Vernon) but owner Larry Passmore has found an unexpected drawback to the location. "We're way up at the north end of the valley," he observes. "By the time the tourists get here, they have a trunk full of wine." The colonial-style winery perches on a hillside with a fine view of the valley and the 15-acre (six-hectare) vineyard that Passmore began planting in 1991. Born in Vernon in 1950, Passmore was running a winemaking supplies store when, with a number of

STORED WITHOUT LABELS, SUMAC RIDGE CLASSICALLY MADE SPARKLING WINES MATURE UNTIL READY FOR MARKET.

local partners, he acquired the winery property. Grapes had flourished there from 1966 until 1988, when they were pulled out. He planted Maréchal Foch, Pinot Noir, Auxerrois, Gewürztraminer and Chardonnay on a warm slope that dips steeply to the south. "I'm happy with the farm and I'm happy with my choice of varieties," Passmore says. His slope, he has found, averages about 230 frost-free days a year, or about 100 days more than the airport at Kelowna. "We have a very, very long day."

The winery, opened in 1994, produces affordable wines largely for the regional market. Passmore has made it a destination visit by building walking trails, picnic areas and a small golf course. In a 100,000-gallon (454,000-litre) tank by the winery, he rears rainbow trout for customers to catch and barbecue. Weddings and concerts often take place here. Irrepressibly informal, Passmore sometimes invites visitors to pitch in at the bottling line. As he says on the winery Web site, "Bella Vista can best be described at this point as a hobby taken to a ridiculous extreme by a few fun-loving wild and crazy guys."

The Bonaparte Bend Winery is just north of Cache Creek, the gateway to British Columbia's Cariboo ranching country. JoAnn Armstrong, who manages this fruit winery, grew up on the storied Gang Ranch north of Clinton, once the world's largest cattle ranch. Her husband, Gary, is a veterinarian who once had interests in five animal hospitals in the interior of British Columbia. In 1980, the couple moved the practice to Cache Creek, buying a 160-acre (65-hectare) ranch on the bank of the Bonaparte River. (A French fur trader named the

THE RECLINE RIDGE VINEYARD NEAR SALMON ARM WAS FORMERLY A PASTURE.

river for Napoleon Bonaparte.) In a rare exception to the rule, this winery, which opened in 2000, did not emerge from an infatuation with wine: the Armstrongs are occasional wine consumers only. "I mix my wines with club soda or put fresh fruit in them," JoAnn admits. She wanted to launch a business at their Bonaparte Ranch, which already had a five-acre (two-hectare) orchard growing apples, raspberries and saskatoon berries. A fruit wine that a guest had brought to a 1998 Christmas party inspired her to write a business plan for a fruit winery. The winery, including a bistro, is in a handsome new building beside the highway. "I could have started in a barn — there is a barn on the property — but I didn't think that was the proper facility for a winery."

Again, the distance from the Okanagan isolated the Armstrongs from their peers. They managed to get one winemaking consultant to make the three-hour drive twice and could not elicit much interest from others. It forced this resourceful couple to take winemaking courses and become self-sufficient. Using both their own fruit and purchased fruit, they produce pleasant wines from apples, blackberries, boysenberries, raspberries, chokecherries, blueberries, rhubarb, cranberries and apricots. Most of the wines finish dry or nearly dry (even the mead made from honey produced on the ranch is not overly sweet) so they can be consumed with main-course dishes. "One customer came in and told us how good the blueberry wine is with marinara sauce," she says with satisfaction.

A two-hour drive east of Osoyoos, Tom Bryden and Lawrence Wallace, his son-in-law, opened the first winery in the Kootenays in 2001. Located 10 miles (16 kilometres) south of Trail, Columbia Gardens and its six-acre (2.4-hectare) vineyard are in a valley beside the swift-flowing Columbia River. Warm and fertile, the district is known locally as Columbia Gardens because both tree fruits and vegetables grow well here. The vineyard is on a southwest-sloping bench above the river, on property the Bryden family has owned since the 1930s. With Bryden working in the purchasing departments of several of Trail's major employers and

THE OLD WINEMAKING PRACTICE OF CRUSHING GRAPES BY FOOT TAKES PLACE AT THE ANNUAL GRAPE STOMP COMPETITION DURING THE FESTIVAL OF GRAPE IN OLIVER.

Wallace busy as a plumbing and heating contractor, the property had become a hobby farm.

Wallace had been a keen home winemaker for almost two decades. In the early 1990s he planted some vines experimentally, quickly zeroing in on varieties suited to this microclimate. The first significant planting was Maréchal Foch, the winery's largest volume red wine. "We like the track record on that variety," Bryden says. "We knew it was pretty hardy." Now Foch has been joined in the vineyard by Pinot Noir, Chardonnay, Auxerrois and Gewürztraminer along with a few small blocks of early-maturing aromatic German whites. "We can grow the same varieties as they can in the Okanagan, but they will taste different," Bryden says. The winery, with a tasting room in an attractive log building, sold its debut wines (less than 1,000 cases) by Christmas 2001. The success did not go unnoticed. The following year, a neighbouring farmer in the Columbia Valley planted 1,500 vines on a property with the potential for a large vineyard. Other farmers are watching with interest. "I think we've broken the ice and proven it can be done," Bryden says. "It could develop into something significant."

The Pemberton Valley Vineyard, which opened in 2000, evolved from the home winemaking experience of Patrick Bradner, a real estate agent born in Vancouver in 1957. In 1987, he and his wife, Heather, moved to an acreage north of Pemberton, a mountain village near the big ski resort at Whistler. After years of making wine from purchased grapes, Bradner planted a three-acre (1.2-hectare) vineyard in 1997. He finds that his vineyard receives at least as much warm sun as those in the Cowichan Valley although he suffers a higher risk of frost in the spring and fall. Bradner started to make wine commercially for his new winery from Okanagan grapes, beginning with the 1999 vintage. At the same time, the Bradners developed a bed and breakfast inn in which each of the suites has a wine name: Burgundy, Bordeaux and Champagne. The guests and the local community have been eager consumers of Pemberton Valley's limited vintages of Chardonnay, Pinot Gris, Maréchal Foch and a Merlot-Cabernet blend. Unhappily for Bradner, his vineyard has been stricken with a wasting disease called crown gall. He is replanting gradually with disease-free vines and fine-tuning the varieties, dropping Chardonnay in favour of an early-ripening Muscat variety.

THE WINERIES

Bella Vista Vineyards
3111 Agnew Road, Vernon, BC, V1H 1A1
Telephone: 250 558-0770
www.webtec.com.au/bvv

Bonaparte Bend Winery
Highway 97, Cache Creek, BC, V0K 1H0
Telephone: 250 457-6667
E-Mail: bbwinery@goldtrail.com

Columbia Gardens Vineyard & Winery
9340 Station Road, Trail, BC, V1R 4W6
Telephone: 250 367-7493
E-Mail: lwallace@netidea.com

Hunting Hawk Vineyards
4758 Gulch Road, Spullumcheen, BC, V0E 1B4
Telephone: 250 546-2164
E-Mail: hunthawk@cnx.net

Larch Hills Winery
110 Timms Road, Salmon Arm, BC, V1E 2P8
Telephone: 250 832-0155
www.larchhillswinery.com

Pemberton Valley Vineyard & Inn
1427 Collins Road, Pemberton, BC, V0N 2L0
Telephone: 604 894-5857
www.whistlerwine.com

Recline Ridge Vineyards & Winery
2640 Skimikin Road, Tappen, BC, V0E 2X0
Telephone: 250 835-2212
www.recline-ridge.bc.ca

OLD VINES COUNTRY—
THE VINEYARDS OF KELOWNA

"THE FIRST TENET OF OUR PHILOSOPHY IS RESPECT THE LAND.
THE KEY TO MAKING PREMIUM WINES STARTS IN THE VINEYARD."
—GORDON FITZPATRICK
CEDARCREEK

Pinot Reach Cellars released two vintages of Riesling before proprietor Susan Dulik was crafty enough to call it Old Vines Riesling, beginning with the 1999 vintage. Now one of her most successful wines, it is so named because the vineyard, on a southwestern slope in east Kelowna overlooking the city and the lake beyond, is one of the oldest commercial vineyards in British Columbia. Mature vines often produce superior wine.

The province's wine industry (excluding the earlier fruit wines of Vancouver Island) began on these slopes in 1925 when J.W. Hughes, a horticulturist from Iowa, began planting both table and wine grapes. By 1944, he had about 300 acres (121 hectares) in vines, primarily North American labrusca varieties such as Bath and Delaware. Hughes sold the best vineyards to his three foremen, one of whom was Czech-born Martin Dulik, Susan's grandfather. The new owners began the transition to better wine grapes. In 1978 the former Jordan & Ste-Michelle Cellars winery in Surrey persuaded Dulik and his son, Daniel (Susan's father), to plant Johannisberg Riesling and other vinifera grapes. The entire 33-acre (13.4-hectare) vineyard was replanted by 1982. The Riesling is the famous clone developed in the Mosel by Hermann Weis and now widely planted around the world. "I recall planting the first vine of Riesling at Dulik's,"

OPPOSITE: **THE 1897 LOG BUILDING, MALLAM HOUSE, IN THE SUMMERHILL VINEYARD IS NOW A MUSEUM.**

VINTAGE TASTINGS ARE GENEROUS AT PINOT REACH CELLARS.

recounts Lloyd Schmidt, a consulting viticulturist whose father also was a Hughes foreman. The vine was planted "with much ceremony" by the Duliks and winery representatives. "I recall toasting the vine with German Riesling," Schmidt says, "then pouring the wine onto that first vine, wishing it and the planting a good future."

These vines, among the oldest vinifera in the north Okanagan, now produce Pinot Reach's Old Vines Riesling. The wine's spicy fruit and full texture reflect the intense flavours of grapes from mature vines. In Europe, a vine at 25 is considered middle-aged but at the full power of its production. Susan Dulik is fortunate to have fruit from such mature vines: most vines in British Columbia are less than 10 years old. (She is not the only winemaker to draw attention to mature vines. Quails' Gate Estate Winery near Westbank has achieved cult status with its Old Vines Foch, made since the 1994 vintage from grapes grown on vines planted in 1969.)

The Dulik vineyard (known from the Hughes era as the Pioneer Vineyard) is near the mission that Father Charles Pandosy, a French Oblate priest, established in 1859. "It is a great valley situated on the left bank of the great Lake Okanagan and rather near the middle of the Lake," he said in a letter written soon after he and his band of missionaries arrived in the valley. "The cultivable land is immense and I myself believe that if Fr. Blanchet [a fellow Oblate in Oregon] is able to send us next year some vine cuttings we shall be able to start a plantation, for when Bro. Surel arrives, if he accepts my plans, we shall elevate our little demesne to the middle of the plain, against a little hill very well exposed...." On the basis of that little vineyard for sacramental wine, Father Pandosy has been called, somewhat extravagantly, the father of the British Columbia wine industry.

Both Father Pandosy and J.W. Hughes knew good vineyard locations when they saw them. "A lot of the original properties were planted to take advantage of the air drainage," notes Roger Wong, Pinot Reach's winemaker. "In the 1920s, they did not have fans to combat the frost." The flatter vineyards farther south at Oliver have installed fans that, powered by propane-fuelled muscle car engines, agitate air over the vineyards. During spring or autumn frosts, the fans prevent cold air from gathering in vegetation-freezing pools in the vineyards. The steep slopes of the east Kelowna vineyards spill the cold air into the valley, away from the vines. Even these vineyards suffered damage, however, when Okanagan winters were colder than they have been in recent years. Half the vines in Pioneer Vineyards were killed by frost in 1964 and again in 1969.

The vineyards supporting the St. Hubertus winery, planted initially by Hughes in 1938 and 1939 when the property was called Lakeside Ranch, are on slopes so steep (grades of 25 to 30 percent) that a former owner was crushed to death when his tractor rolled over. The movement of air down the slopes combats frost and the lake moderates late-season temperatures. That, Wong says, is why Gray

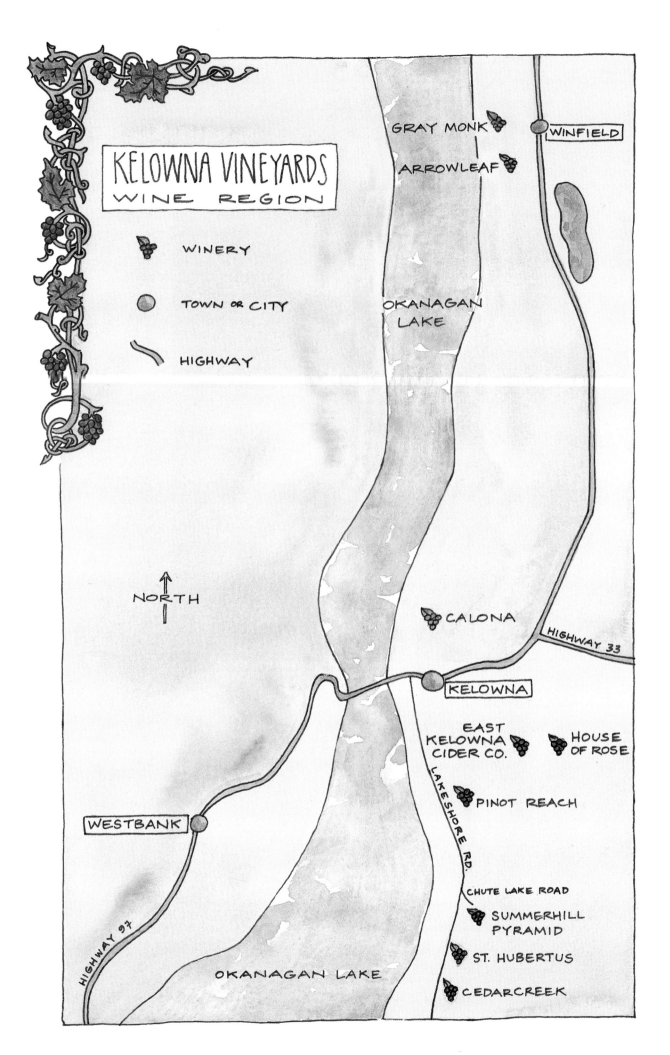

GORDON FITZPATRICK
CEDARCREEK ESTATE WINERY

SUSAN DULIK
PINOT REACH CELLARS

KELLY MOSS AND HOWARD SOON
CALONA VINEYARDS

GEORGE HEISS JR.
GRAY MONK ESTATE WINERY

Monk, whose vineyard at Okanagan Centre tips westward toward the lake, succeeds at one of the valley's most northern sites. Other Kelowna wineries with vines perched on slopes admired by Father Pandosy include CedarCreek, Summerhill and Arrowleaf. Calona Vineyards, with its winery in Kelowna, owns vineyards south of Oliver but has bought Kelowna-area grapes since 1935.

Unlike Hughes and Father Pandosy, Virgil and Eugene Rittich, brothers who grew up in a family vineyard in Hungary, did not select good sites when they began growing grapes in the Kelowna area in 1930. Eugene was a graduate in viticulture as well as law and Eugene was a mechanical engineer. They came to Canada after World War I and Eugene became the winemaker for Growers' Wines in Victoria. Together, they planted grapes north of Kelowna, a project they described in British Columbia's first wine book, *European Grape Growing in cooler districts where winter protection is necessary*, published in 1941. "We began our work in 1930 and planted in the same year the first grapes in the Black Mountain district 10 miles [north of] Kelowna, at an altitude of 2,000 feet above sea level," they wrote. "The place was not ideal having cold winter temperatures (in two years -30°F) and very cool nights at blossoming time, sometimes only a few degrees above the freezing point."

The Rittich vineyard was not far from the north end of today's Kelowna airport, away from the lake and marginal for grapes. (The first Hughes vineyard, supplying an ultimately unsuccessful grape juice and jelly business, also was in this area.) Had the Rittich brothers been on one of the better Hughes vineyard sites, the history of British Columbia wine might have been different. They tested about 40 European grape varieties, finding

AT OLIVER'S OKANAGAN BARREL WORKS, CALVIN CRAIK BUILDS, REPAIRS OR IMPORTS PREMIUM OAK BARRELS FOR BRITISH COLUMBIA WINERIES.

a number promising, including Pinot Blanc and Pinot Noir. "More than 100 farmers have tried to grow our varieties," they wrote. "All who took care of them had success and corroborated our results." Yet none of those farmers adopted the vinifera, perhaps because the wineries paid the same for easy-to-grow labrusca grapes as they offered for

WINERY OWNER VERN ROSE PROUDLY OFFERS HIS WINES AT FESTIVALS.

vinifera, which were difficult to grow. After Eugene Rittich died, his widow had the vinifera replaced with hardier hybrid grapes in 1962.

Today, House of Rose is the only Kelowna-area winery with a vineyard not benefiting from the lake effect. The winery was established in 1993 by Vern Rose, a former Alberta schoolteacher who had retired five years earlier to a seven-acre (2.8-hectare) farm near Rutland, a northern suburb of Kelowna. When he got interested in growing wine, Rose shrewdly went to New Zealand to learn something about cool-climate viticulture. He recognized that his site is one of the coolest in the Okanagan; the Merlot vines already there when he bought the farm only ripen adequately half the time. However, House of Rose does benefit from good air movement — an almost daily wind called the Belgo Breeze — that wards off frost. Rose also planted varieties more appropriate for his site.

The vineyards of St. Hubertus, Summerhill and CedarCreek lie on the same bench as Pinot Reach. The Riesling at St. Hubertus also dates from 1978, although Leo and Andy Gebert, the owners, do not offer an "old vines" label. Leo Gebert, who trained as a banker in Switzerland where he was born in 1958, bought the 55-acre (22.3-hectare) vineyard in 1984. His younger brother Andy, who emigrated from Switzerland five years later, took over about half the vineyard and became a partner in the winery, which opened in 1992. This vineyard's history mirrors the evolution of British Columbia wine growing. As Lakeside Ranch, it grew labrusca and table grapes. Frank Schmidt acquired it in 1958 and planted better winemaking grapes. His most curious choice was a variety called Himrod, a seedless table grape developed in New York State. The general manager at Growers' Winery was ecstatic about the quality of the grapes, begging Schmidt to keep his plantings secret and not sell cuttings to anyone else. "Himrod table wine may yet make you famous," Schmidt was told. Nothing of the sort happened. The winery released the Himrod wine as Canadian Liebfraumilch and, when the German wine industry threatened to sue, changed the name to Rhine Castle. Schmidt later sold hundreds of cuttings to hobby gardeners across western Canada. One of the original Himrod vines, now with a massive trunk and heavy with succulent grapes each summer, still thrives outside the modest St. Hubertus tasting room.

Leo Gebert took over the vineyard just in time to experience one of the Okanagan's coldest winters. "Nineteen eighty-five was brutal," he recalls. "At the beginning of November, it fell as low as minus 28 and it stayed cold for five weeks. The first cheques I got were from crop insurance. It was a little bit devastating." Yet both the Riesling and young Pinot Blanc vines planted that spring survived the winter, encouraging Gebert to continue replanting the vineyard. The brothers have succeeded with varieties such as Gewürztraminer, Chardonnay, Gamay, Pinot Meunier and Maréchal Foch, along with a sentimental favourite, Chasselas, the most widely planted white grape in Switzerland. "If I weren't Swiss, I wouldn't plant it again," Leo Gebert grouses. "It is so easily damaged by a late spring frost." St. Hubertus and Quails' Gate, whose vineyards on the west side of Okanagan Lake are directly across from St. Hubertus, are the only two wineries in British Columbia even producing Chasselas. The original Chasselas was planted at Quails' Gate in 1963. Richard Stewart, who had begun developing that vineyard in the mid-1950s, had ordered White Diamond, a labrusca variety, from an American nursery, only to be shipped a superior vinifera grape in error. More was planted in 1975 and again in the 1980s. Chasselas yields fruity and inexpensive whites. Both St. Hubertus and Quails' Gate have strong followings for their wines.

The wineries neighbouring St. Hubertus, Summerhill to the north and CedarCreek to the south, are on slopes similar to those that caught the eye of J.W. Hughes. The Summerhill property was acquired in 1986 by Stephen Cipes, a former real estate developer from New York who came to the Okanagan for a life closer to nature. He grows grapes organically on about 40 acres (16 hectares), where the varieties include Riesling vines almost as mature as those of his neighbours. The winery has converted many of its growers to organic techniques as well. Behind this thrust are deeply held environmental views. "I'm 58 years old," Cipes said in 2002. "I grew up as a boy on Long Island Sound and went fishing there with my dad. By the time I left New York in 1986, there were no fish left. You could see garbage floating in the waters of the Sound. When I came here and saw how people treated this lake, I was appalled. Thousands of tons of chemicals from agriculture drain into the lake.

SPERLING VINEYARD OVERLOOKS THE OKANAGAN VALLEY AT KELOWNA.

STEPHEN CIPES
SUMMERHILL PYRAMID WINERY

ANDY AND LEO GEBERT
ST. HUBERTUS ESTATE WINERY

While vineyards have a lot less need of chemicals than orchards, the first thing I did was eliminate pesticides and herbicides."

Since the winery opened in 1992, the signature products have been sparkling wines. "Flawless sparkling wine can only be made with organic grapes," Cipes maintains. "That's my opinion. If you have the slightest chemical taste in the grapes, it will show in the wine." The grapes grown near the winery are primarily for sparkling wine: Riesling for Cipes Brut; and Pinot Noir, Pinot Meunier and Chardonnay — the classic trio of Champagne — for Summerhill's other sparkling wines. None of the wines has been certified as organic; Cipes thinks this might be viewed negatively in some markets. "But I know I am doing my duty as a citizen of the planet by growing my grapes organically."

An individual with original opinions, Cipes ages Summerhill's wines, both sparkling and still, in the cool silence of the white pyramid uphill from the large winery. In its first years, when Summerhill was making wines in a three-car garage, the winery built a small pyramid to test Cipes's theories. In daily tastings over three years, consumers were asked to compare pyramid-aged wines with similar wines that had never been in the pyramid. "The tasters chose the pyramid-aged wine almost unanimously every day as being smoother and having a better aroma," Cipes recounts on the winery's Web site. When the current winery was built (with a restaurant and one of the biggest tasting rooms in the industry), Cipes had a large pyramid, an eight percent scale model of Egypt's Great Pyramid, erected uphill from the winery. There are, to be sure, plenty of sceptics about the impact of the pyramid's "life-force" on the wines. The rejoinder from Cipes is that 18 of the 22 wines made by Summerhill in 2000 won acclaim in competitions. In 2002, Cipes even changed the winery's name to Summerhill Pyramid Winery.

In contrast with East Kelowna's other winery vineyards, most of the vines currently growing at CedarCreek are less than 10 years old. These slopes were planted with fruit trees in the 1940s. The modest vineyard that existed in 1978, when the farm was purchased to develop what became the Uniacke winery four years later, was planted to Okanagan Riesling and De Chaunac. Mediocre wine grapes at best, these varieties were almost totally eliminated in the Okanagan after the 1988 harvest, when two-thirds of the vineyards were pulled out in a government-funded effort to get rid of poor varieties. David Mitchell, Uniacke's owner, planted some vinifera as soon as he took over.

In 1984 Uniacke released a 1981 Merlot, the first varietal from those grapes to be released in British

Columbia. The winery was acquired in 1986 by Senator Ross Fitzpatrick and renamed CedarCreek. Merlot has remained in CedarCreek's portfolio, but not as "old vines" because most of the grapes come from younger vineyards elsewhere in the Okanagan. Much of the vineyard at CedarCreek has been replanted since 1991, some parts of it twice to establish the varieties that make the best wines here. The flagship wines found now in the pristine white Mediterranean-style winery are Burgundian: Pinot Noir, Pinot Gris and Chardonnay. Like most northern wineries, CedarCreek also sources grapes from elsewhere in the Okanagan.

Calona Vineyards, whose historic winery in downtown Kelowna is British Columbia's oldest continually producing winery, has a rich history. In 1932, Guiseppe Ghezzi, a debonair winemaker from Italy, marshalled investors among British Columbia's immigrant Italians to make wine from surplus apples and later from grapes. When the project ran short of money, the founders secured capital from Pasquale "Cap" Capozzi, a Kelowna grocer. He came to be credited as the winery founder because he raised cash by selling shares to other immigrants. He was helped by a Kelowna hardware merchant, W.A.C. Bennett, who served as winery president even though he was a teetotaller. Bennett, who was British Columbia premier from 1952 to 1972, resigned when he entered politics in 1941. Capozzi remained in the background, this time recruiting a local car dealer named Jack Ladd as president. Ghezzi returned to Italy to marry an opera star, leaving day-to-day management in the

A WINE TOUR GROUP STOPS TO ASK QUESTIONS IN CALONA VINEYARD'S VAST CELLAR.

hands of his son, Carlo. When Carlo retired in 1960, Capozzi's three sons — Joe, Tom and Herb — took over. They ran it with élan, nearly enlisting the Gallo family of California as partners before selling to Standard Brands of Montreal in 1971. During the next three decades, Calona had several owners, including a notorious Vancouver stock promoter, before coming under the management of Cascadia Brands, a holding company controlled by a Swiss banker.

Throughout its history, Calona's wines reflected whatever the current taste was, whether it was for sweet red wines, high-alcohol fruit wines or simulations of European table wines with brands such as Schloss Laderheim and Sommet Rouge. To its credit,

THE OKANAGAN LAKE REFLECTS SUN AND WARMTH ONTO THE VINEYARDS AT GRAY MONK.

Calona, under the Capozzi brothers, invested in vineyards south of Oliver in the 1960s. Standard Brands sold the vineyards but in 1997, under Cascadia, Calona acquired an interest in those same vineyards when it invested to help start the Burrowing Owl Estate Winery. When that partnership was dissolved in 2002, Calona remained the owner of 174 acres (70 hectares) of premium vineyard on Black Sage Road.

Those grapes support a virtual second winery at Calona called Sandhill. Production, currently 12,000 cases a year, is directed by Howard Soon. Now 51, the Vancouver-born Soon has worked at Calona since 1980, emerging as the winemaking star with Sandhill, while Kelly Moss, the assistant winemaker, has increased responsibility for wines under the Calona label. Sandhill was launched with the 1997 vintage when Soon began receiving significant

volumes of grapes from Burrowing Owl vineyard. Sandhill makes only vineyard-designated wines. "It is the ultimate that we can do," Soon says of his wines.

Under its Calona label, the winery makes the world's only varietal from a grape called Sovereign Opal. A spicy and floral white, the variety was developed by scientists at the Summerland research station who crossed Muscat and Maréchal Foch. When it was released in 1976, the only grower who embraced it was the late August Casorso, a member of perhaps the oldest wine-growing family in the Okanagan. Ancestor Giovanni Casorso, who came from Italy in 1883 as an agriculturist for the Oblate mission, was one of first backers of Calona founder Guiseppe Ghezzi. Today, John Casorso, August's son, grows Sovereign Opal in a 10-acre (four-hectare) vineyard in east Kelowna.

At the East Kelowna Cider Company, whose tasting room is near the Pinot Reach winery, David and Theressa Ross ferment apples, not grapes. A logger, David Ross, 34, has been fermenting fruit at home since he was 12. He grew up on a 20-acre (eight-hectare) apple orchard, in his family since it was acquired in 1941 by his grandfather, Charles, an immigrant from Romania. The orchard was subsequently split between two brothers. David and Theressa took over one of those parcels in 2002, opening the cidery in order to make apple-growing viable. During the previous five years, they had successfully developed sales of carbonated soft apple cider. "My husband's dream has been to make alcoholic cider, not soft cider," Theressa Ross says.

Theirs is not English cider. "We weren't fond of it," Theressa says. "We wanted to produce something we considered more drinkable." As well, they grow only dessert apples — Red and Golden Delicious, McIntosh and Spartan — and this also defines the style. The juice from these four varieties is blended prior to fermentation and to a formula the Rosses prefer to keep to themselves. Unlike many English ciders, theirs is carbonated, with only six percent alcohol and with some fresh apple juice in the final blend to lift the aroma and the flavours. The cider is packaged in clear bottles, similar to beer.

With limited resources, the Rosses have assembled basic but effective processing equipment. The apples are crushed with a simple hand crusher. The crushed pulp is bagged in nylon sacks, which are placed in household washing machines (there are six). It takes about two minutes in the spin cycle to separate the juice from the residue, with the juice

OKANAGAN WINES ARE OFTEN TASTED IN THE VINEYARDS WHERE THEY GROW.

then allowed to settle and clear before fermentation. "Our equipment is old-fashioned but it works," Theressa Ross says. She can process a bin (800 pounds/363 kilograms) of apples in half an hour.

About half an hour's drive north of Kelowna, the picturesque Gray Monk winery opened in 1982 on a superb southwestern slope that catches the reflected warmth of the sun from the nearby lake during long afternoons and evenings. This definitely is a site for cool-climate viticulture: it shares with Germany's Geisenheim Institute the distinction of being almost on the 50th parallel (a location that inspired Latitude 50, Gray Monk's best-selling white wine). George Heiss, the owner along with wife, Trudy, and two sons, says the vineyard has never had serious damage from either early spring or late fall frosts. Some of his vines meet the Okanagan

definition of "old" because in 1976 Heiss planted Pinot Gris and Auxerrois vines imported from Alsace in France. These were among the earliest successful introductions of classic grape varieties into the Okanagan.

Heiss is an unlikely wine grower. Born in Vienna in 1939, he took up the trade of his father and grandfather, both of whom had been European hairdressing champions. Later, in Edmonton, where he and his German-born wife both had hair salons, clients of the talented George Heiss booked him three months in advance. But after 25 years in the trade, he was ready for a change in 1972 when his father-in-law, Hugo Peter, told him of orchard property that had just come up for sale. "I'd seen too many old hairdressers," Heiss says, "and I didn't want to be one of them." Peter was already living in the Okanagan, having planted Maréchal Foch in 1969 on a vineyard now operated by his grandson, George Heiss Jr., Gray Monk's German-trained winemaker. Now replanted, that vineyard is a source of Gamay for the winery.

"The only thing I knew about grapes and wine was how to drink it," Heiss Sr. says. There is good wine in Vienna, one of the few world capitals with vineyards and wineries within the city limits. Observing the thriving tree fruits on his new Okanagan farm, he calculated that grapes would thrive. The inexperienced Heiss initially planted Foch and another French hybrid called Rosette but soon found them to be second-rate. "They were not exported from France, they were deported," he fumes. He read European literature on grapes and, when his father-in-law went to Germany on a vacation in 1975, Heiss had him bring back vines from the Alsace research station at Colmar to start replanting the vineyard in the spring of 1976. The next year, Helmut Becker, the legendary director of vine research at Geisenheim in Germany, first visited the Okanagan, spending a day with Heiss. "He said, 'You can grow anything' and offered me 34 varieties to try out," Heiss recalls. Heiss countered that the entire industry should share in the trials. The result was the Becker Project, during which a large number of European grapes were grown for

PRUNING DORMANT VINES — THE COLDEST JOB IN THE VINEYARD.

OPPOSITE: **PROTECTIVE SLEEVES SHIELD YOUNG VINES FROM BOTH WIND AND RODENT DAMAGE.**

eight years in two Okanagan test plots. Varieties that succeeded in these trials, notably Pinot Blanc, are now part of the backbone of quality wine in British Columbia. The Becker Project grape that intrigued Heiss was a variety called Rotberger, a cross of Trollinger and Riesling that produces a true rosé wine. Like Calona with its Sovereign Opal, Gray Monk is the only winery in North America making Rotberger, a crisply fruity wine as fresh as a spring morning. Only 1,300 cases of Rotberger are made each year and it sells quickly. The Gray Monk vineyard was planted primarily with Auxerrois, Pinot Gris, Bacchus, Kerner and Gewürztraminer, white varieties well suited to the north Okanagan. For reds, the winery buys from south Okanagan growers.

If Gray Monk has had a disadvantage, it is that the winery, even if only 10 minutes from the highway, seemed the only one for miles around. That changes with the opening in 2003 of Arrowleaf Cellars, a winery almost within walking distance of Gray Monk. The winery's 16-acre (6.5-hectare) vineyard sold grapes to Gray Monk until Arrowleaf's initial crush in 2001. This winery, named after a common plant sometimes called the Okanagan sunflower because of its bright yellow blossoms, is owned by Josef and Margrit Zuppiger and their winemaking son, Manuel. They came to Canada in 1986 from their native Switzerland. "We were just born farmers," says Josef, who was born in 1950. "Here in Canada we could afford a large farm." They operated a dairy farm near Edmonton, but when their sons showed little interest in dairying, they bought the Camp Road vineyard in 1997. With orchard experience in Switzerland, Josef figured he would have no problem growing grapes. Manuel, born in 1976, returned to Switzerland to learn winemaking at Wädenswil research station, polishing his lessons with practical experience at an Australian winery and at two Okanagan wineries.

The two-storey Arrowleaf winery is a little higher on Camp Road than Gray Monk but has a comparably fine view across the lake from its tasting room. The vineyard, with a west-facing slope of 15 to 18 percent, benefits from both the long afternoon and evening sun and the reflection of warmth from the lake. The Zuppigers inherited plantings of Bacchus, Gewürztraminer, Pinot Gris, Dunkelfelder and Auxerrois, some of which were planted in 1986. They added Merlot in 1998, Zweigelt in 1999 and Vidal (for icewine) in 2000. Manuel expresses some disappointment that there is no room for Pinot Noir "but we expect the Zweigelt to do very well." Manuel's recent training expresses itself in winemaking that leans toward crisply dry whites and reds aged in French oak. His father cautions: "Manuel prefers dry wines but some consumers like sweeter wines."

THE WINERIES

Arrowleaf Cellars
1574 Camp Road, Lake Country, BC, V4V 1K1
Telephone: 250 766-5033
E-Mail: jozu@cablelan.net

Calona Vineyards
1125 Richter Street, Kelowna, BC, V1Y 2K6
Telephone: 250 762-3332
www.calona.kelowna.com

CedarCreek Estate Winery
5445 Lakeside Road, Kelowna, BC, V1W 4S5
Telephone: 250 764-8866
www.cedarcreek.bc.ca

East Kelowna Cider Company
2960 McCulloch Road, Kelowna, BC, V1W 4A5
Telephone: 250 860-8118
E-Mail: ekcider@attcanada.ca

Gray Monk Estate Winery
1055 Camp Road, Okanagan Centre, BC, V4V 2H4
Telephone: 250 766-3168
www.graymonk.com

House of Rose
2270 Garner Road, Kelowna, BC, V1P 1E2
Telephone: 250 765-0802
E-Mail: arose@shuswap.net

Tantalus Vineyards
1670 Dehart Road, Kelowna, BC, V1W 4N6
Telephone: 250 764-0078

St. Hubertus Estate Winery
5225 Lakeshore Road, Kelowna, BC, V1W 4J1
Telephone: 250 764-7888
www.st-hubertus.bc.ca

Summerhill Pyramid Winery
4870 Chute Lake Road, Kelowna, BC, V1W 4M3
Telephone: 250 764-8000
www.summerhill.bc.ca

THE VINEYARDS OF MOUNT BOUCHERIE

"I DON'T KNOW HOW GOOD WE CAN BE. IT IS TOO EARLY TO TELL BECAUSE THE VINEYARDS ARE TOO YOUNG. BUT THE POTENTIAL OF BRITISH COLUMBIA HASN'T BEEN BOTTLED YET."

—JOHN SIMES
MISSION HILL FAMILY ESTATE

Rising from the floor of the Okanagan valley, Mount Boucherie stands alone, a geological stray from the valley's mountainous western boundary. Some of the best vineyards of the north Okanagan are on the mountain's southeastern flank, where the vines are in sunlight from early morning until late in the day. Housing estates have also invaded this slope, which is perhaps why there only are four wineries here. They are, however, wineries of individuality. The most notable is the stunningly rebuilt Mission Hill Family Estate winery. As elegant as a Tuscan hilltop palace, it is the Okanagan's most visited winery.

The first significant vineyards were planted on the slope by Richard Stewart, whose sons, Ben and Tony, now operate Quails' Gate Estate Winery. A member of a prominent Okanagan farming family, Stewart was something of a pioneer in 1956 when he bought what today includes the 85-acre (34.4-hectare) vineyard at Quails' Gate. This was two years before the floating bridge was completed across the lake at Kelowna, opening up subsequent commercial and residential development around Mount Boucherie. Stewart's property was then known as the Allison Ranch. In 1873 a cattleman named John Allison sought to raise livestock here but several hard winters devastated the herd. The cabin that Allison and his wife, Susan, built has remained. In 1989, when Quails' Gate winery opened,

OPPOSITE: **VINEYARD AT QUAILS' GATE WITH PICTURESQUE OKANAGAN LAKE AS A BACKDROP.**

the Stewart family restored the cabin as the winery's handsomely rustic tasting room.

There was little expertise in Pacific Northwest viticulture when Stewart began planting vines in 1963. Conventional wisdom held that European grape varieties were no more likely to survive than John Allison's cows. Dealing with a nursery in Washington State, Stewart ordered a selection of North American vines, such as Campbell's Early, Sheridan, Patricia and White Diamond, none of which are acceptable today for winemaking. The nursery mistakenly substituted Chasselas, an important European vinifera in Swiss vineyards, for Diamond. After a visiting European grape expert pointed out the mistake a few years later, Stewart happily planted more Chasselas. He pulled out the North American vines, replacing them first with French hybrids and ultimately with classic vinifera such as Pinot Noir, Chardonnay and Riesling. These are now the signature wines at Quails' Gate but, to this day, the mature Chasselas vines produce a crisply fruity white wine that sells out vintage after vintage.

Before Richard Stewart planted grapes, the slope of Mount Boucherie was given over to orchards, one of which was purchased in 1962 by Joe Slamka. A Czech-born machinist who left his native land when the Communists took power, Slamka (the name means "straw" in Czech) was working in an Edmonton pipe mill when he was attracted to the Okanagan during a camping vacation. Beginning in 1979, he converted the orchard to a nine-acre (3.6-hectare) vineyard. His son Peter took over the property in 1989 and opened the Slamka Cellars winery in 1996. In 2000 Peter Slamka (with a brother) acquired a neighbouring orchard, turning that into vineyards as well.

The combined vineyards of the Slamka family, totalling about 23 acres (9.3 hectares), have an array of vines, including one of the Okanagan's few remaining plantings of Seyval Blanc. This is a vigorous white French hybrid that yields both table wine and icewine. The vineyard's aromatic whites, Schönburger, Siegerrebe and Traminer, comprise a popular blend the winery calls Tapestry. A favourite variety in the vineyard is Auxerrois; some were planted here during the Becker Project and are now the second-oldest Auxerrois in the Okanagan (after Gray Monk). The red varieties include Pinot Noir, which grows well on the slope of Mount Boucherie. Slamka also grows Lemberger, Merlot and Maréchal Foch. "I'm going to make a wine labelled Big Red," Peter Slamka promises.

In 1958, wheat farmer Mehtab Gidda, whose brother already worked in a British Columbia sawmill, emigrated from the Punjab in India. He lived near Kamloops for two years, but when that proved too rugged for his wife he bought a small orchard on the Boucherie slope. "He is a very hard worker," his son Sirwan Gidda says. For years, the elder Gidda, who had the equivalent of a fourth-grade education, worked at sawmills and farmed in his spare time, financing his family's schooling. "Dad wouldn't let us get into farming until we got our education," says Sirwan. He has a business administration diploma; brother Nirmal has a bachelor of science degree and Kaldeep, the third brother, is a mechanical tradesman. Among them, they operate three major vineyards totalling 170 acres (69 hectares) and the Mt. Boucherie winery, which opened in 2001. "My kids and my brothers' kids — they're not going to be farmers," Sirwan says. "They'll be in the wine business."

The family purchased its first Boucherie-area vineyard in 1975 and actually planned to replace the vines with apple trees but the price of apples collapsed. The family prospered with grapes, however, because there was a guaranteed market, at least until 1988 when two-thirds of the vines in Okanagan vineyards were pulled out. The Gidda family not only did not pull out vines but went looking for more property. They considered the former Monashee Vineyard on Black Sage Road but instead, in 1991, bought an Okanagan Falls vineyard that included a house; Kaldeep's house had burned down the year before and he needed one. In 1997, they purchased the 89-acre (36-hectare) Similkameen Vineyards in the Similkameen Valley. With their various vineyards, the Gidda brothers may well be growing more grape varieties than any other grower. The most unusual is Michurnitz, a full-bodied red

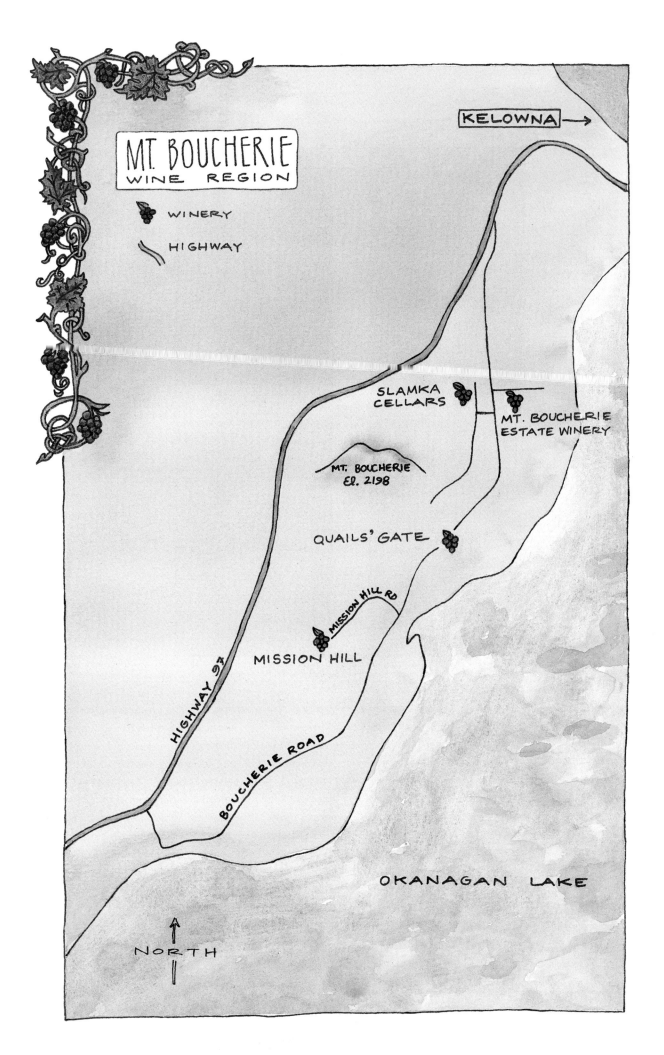

ANTHONY VON MANDL
MISSION HILL FAMILY ESTATE WINERY

ALAN MARKS
MT. BOUCHERIE ESTATE WINERY

PETER SLAMKA
SLAMKA CELLARS

BEN AND TONY STEWART
QUAILS' GATE ESTATE WINERY

variety originally from Russia. "Every time we see something we like, we go for it," Sirwan says. The winery was built because the shrewd brothers could see a surplus of grapes coming in the Okanagan, threatening the prices of the 600 tons (545 tonnes) of grapes they grow unless they added value at a winery. To make their wine, they recruited Summerhill's former winemaker, Alan Marks, a Missouri native with a doctorate in the chemistry of sparkling wines.

The Mission Hill winery, now the grandest of the Boucherie wineries, opened in 1966, perched on the mountain at an elevation of 1,558 feet (475 metres), or 426 feet (130 metres) above the surface of Okanagan Lake. The original investors in Mission Hill (one of whom was a relative of Richard Stewart) were led by entrepreneur R.P. Walrod, who earlier had founded the Sun-Rype apple processing firm. Familiarly known as Tiny because he was large, he conceived a winery in the style of the Napa Valley. Unfortunately, he suffered a fatal heart attack before the winery, resembling a California mission, was completed. The winery's commanding location afforded a fine view from the tasting room, but this was a decade before winery tasting rooms were permitted in British Columbia and the view was wasted. In spite of its location, the original Mission Hill winery slipped into receivership twice during its first 15 years. However, Walrod's vision has not only been realized but lavishly surpassed by Anthony von Mandl, who has owned Mission Hill since 1981. The winery has been rebuilt entirely in a timeless European style that includes Greek arches, Roman colonnades and a slender bell tower spiking 120 feet (37 metres) into the sky. The open-air restaurant and an amphitheatre for concerts look out over the valley and the lake. Vast cellars have been dug into the mountain. The cellar open for public tours has the ambiance of a cathedral. One of Canada's most striking wineries, Mission Hill attracts more than 150,000 visitors to the top of Mount Boucherie each year.

A demanding perfectionist, von Mandl was barely of legal drinking age in 1972 when he became an independent wine merchant in Vancouver, where he was born. His Vienna-born father, whose textile mills in Czechoslovakia had been taken over by the government, set out to rebuild the family's prosperity in Canada and ultimately was affluent enough to have his son educated in private schools in Europe. There, he became close to a Mosel wine producer called Josef Milz who fired von Mandl's interest in wine and later backed the young man in starting a wine agency. Late in the 1970s, Milz commissioned von Mandl to research a possible Milz winery investment in the Okanagan. Milz decided against it but von Mandl proceeded on his own. When he acquired Mission Hill, then trading under the name of Golden Valley Winery, it was a virtually bankrupt winery with a dirt floor and little palatable wine in its tanks. Against all the odds, von Mandl declaimed a vision of an Okanagan filled with quality vineyards and wineries and made it happen.

The key was hiring, in the summer of 1992, the well-trained and talented John Simes, then chief

NEW FRENCH OAK BARRELS MATURE PREMIUM WINES IN THE QUAILS' GATE WINERY.

winemaker at New Zealand's largest winery. Simes, whose wife is Canadian, knew he had made the right move as soon as he walked through the vineyards of nearly mature Okanagan Chardonnay. "This fruit was spectacular," he recalls. "I was absolutely blown away by the quality of the grapes." At his request, Mission Hill bought 100 new oak barrels and Simes crafted a Chardonnay that won the top trophy at a London wine competition early in 1994. This award, well publicized by Mission Hill, gained international credibility (for the first time) for British Columbia wines and gave tremendous momentum to Mission Hill. Today Simes has more than 6,500 barrels and about 1,000 acres (405 hectares) of winery-owned vineyard to feed those barrels. Most of the vineyards are in the very south of the Okanagan. Simes continues to produce some of the Okanagan's best Chardonnay, while gaining renown for reds with grapes from the newer vineyards. Some of Mission Hill's best Bordeaux reds are blended into a premium wine called Oculus. The road up the Boucherie slope to the winery runs by a young Riesling vineyard developed in the late 1990s. From these grapes, Simes makes wines as brilliantly vibrant as the Mosel wines von Mandl once sold.

After seeing how dramatically Mission Hill's wines improved with a winemaker from the southern hemisphere, the Stewart brothers at Quails' Gate hired the first of their several Australian winemakers in 1994. Jeff Martin (who now operates his own La Frenz winery near Naramata), had started in 1977, at the age of 20, as a trainee at the big family-owned McWilliams winery at Griffith, his home

PREVIOUS PAGES: **WINE MATURES IN THE RESTFUL REVERENCE OF MISSION HILL'S COOL UNDERGROUND CELLAR.**

THE PASTORAL OKANAGAN VALLEY FROM THE SLOPES OF MOUNT BOUCHERIE.

town in Australia. By 1989, he was the chief winemaker at the McWilliams premium winery. In British Columbia, Martin stamped his Australian style — fruit-driven and often boldly oaked — onto the Quails' Gate wines. Noting that the oldest red vines on the property were Maréchal Foch, he set out to make a wine as intense as an Australian Shiraz from old vines. With many old vinifera vines, the yield per vine declines naturally as the vines age; they compensate by yielding richly concentrated wines. French hybrid varieties like Foch produce prodigiously if not controlled, resulting in thin, acidic wines. In Canadian vineyards, Foch was an over-cropped workhorse. At Martin's suggestion, Quails' Gate thinned the number of bunches per vine, reducing the yield by about 60 percent from the historic average. The remaining bunches ripened with succulent flavours, enabling Martin to make a full-bodied red tasting sweetly of dark plums. With the 1994 Old Vines Foch, Quails' Gate saved Foch from otherwise assured obscurity.

Martin left Quails' Gate after the 1998 vintage, to be replaced by another Australian, Peter Draper, who had been making fine Pinot Noir at a boutique winery there. Sadly, Draper, only 39, died in November 1999, midway through his first vintage at Quails' Gate. Once again, the winery returned to Australia (after briefly wooing a Canadian working in Washington State), this time for Ashley Hooper. The strapping football-playing son of a veterinarian, Hooper was born in 1969 in the Goulburn Valley near Melbourne. After graduating from Roseworthy College, Australia's renowned winemaking school, Hooper bounced around wineries, ending up at Cellarmaster Wines, a large Barossa winery. He stayed there until he learned that the job had come open at Quails' Gate. "I'd always wanted to go to California; I wanted to work in a small premium winery in North America but I didn't really think I'd achieve it," he says. "Most people [in Australia] think Canada is full of snow that never melts." Hooper arrived at Quails' Gate in April 2000, settling into the job just as Quails' Gate was expanding the winery to an annual capacity of 40,000 cases.

"What I like about it here is that it is a new and emerging wine industry," Hooper continues. "Every wine I make is a whole new project." In particular, he had never before made Pinot Noir, the backbone of the Quails' Gate portfolio. Hooper took a quick, three-day trip to Oregon, visiting a dozen wineries and garnering tips on Pinot Noir. "They shared a lot of information about how they made Pinot," he recalls. "That was invaluable to me. Without that, I would not have made the quality of Pinot that we made in 2000. That cured that little glitch in my experience." Hooper did not just settle into his job; he took it by storm, learning the new vineyards and the new winery while imposing his approach brusquely. "I was going like a bull out of the gate," he admits. "I told Ben Stewart when I took this job on, 'I'm not coming here just to make your Chardonnay and your Pinot. I'm not going to be involved with 19 different wines and only think two of them are great. I'm going to make everything great.'" Initially, he would not have won a popularity contest. One cellar worker walked out over Hooper's abrupt style and others came close to it. He figured he did not have time to explain matters he considered urgent until the new vintage was under control. "The [quality of the] 2000 vintage was above my expectations," he said later. Having met his self-imposed test of fire, Hooper has relaxed somewhat, confident in himself and his crew. "As long as you don't get to the point where you think you've made it. I'm very comfortable with our 2000 and 2001 vintages but that does not mean I will be happy with that for the rest of my life. I'm expecting to do better this year and next year."

Like Mission Hill's Simes, Hooper has been impressed with the quality of the fruit Okanagan vineyards are delivering. "That's what I like about being here — this broad range of interesting wines that we make," Hooper says. "The quality is very good. The flavours and characters we get out of this growing area are really good." Quails' Gate buys grapes from several different regions in the Okanagan but relies on its own vineyards for much of its harvest. Hooper maintains that the winery's vineyards on the slopes of Mount Boucherie are producing the best grapes for him. "There is minimal rainfall and minimal disease here and long days of sunshine," he notes. It is, he adds, a "winemaker's dream" that large blocks of the vineyards, which are immediately adjacent to the winery, have almost entirely been replanted in recent years with the best French clones of such key varieties as Pinot Noir, Chardonnay and Pinot Gris. He recognizes that he "walked in at the right time, when a lot of these vineyards are just starting to kick in. Previous winemakers haven't had the resources I've had."

The disciplined growing that Jeff Martin first imposed in the Foch vineyard now extends to all the vineyards growing for Quails' Gate. The winery produces 60 percent of its grapes from its Mount Boucherie vineyards and from a small one near Osoyoos. The rest comes from growers in various Okanagan locations. "I can't afford that fruit to be ordinary," Hooper says. "I want quality, not quantity."

In 2001, Hooper began negotiating contracts to pay growers a total price for their grapes rather than, as has been traditional, paying by the ton. That gives Hooper the freedom to ask for whatever yield he believes is needed in any particular vintage. "If I get two tons an acre, that's my fault if I told them to chop it off," the winemaker says. "I know of a grower for another winery who said, 'They pay me less but I just grow more tons.' He just does the game. I don't want to do that with growers. I want to work on the vineyards, because you can't make honey out of ducks. You've got to have good grapes." At Quails' Gate, he has a well-equipped winery and all the tools he needs. "If we get good grapes, we're going to make good wine."

OPPOSITE: **THE FERMENTATION TANKS AT QUAILS' GATE ARE PREPARED FOR THE VINTAGE.**

THE WINERIES

**Little Straw Vineyards
(formerly Slamka Cellars)**
2815 Ourtoland Road, Kelowna, BC, V1Z 2H5
Telephone: 250 769-0404
www.littlestraw.bc.ca

Mission Hill Family Estate Winery
1730 Mission Hill Road, Westbank, BC, V4T 2E4
Telephone: 250 768-7611
www.missionhillwinery.com

Mt. Boucherie Estate Winery
829 Douglas Road, Kelowna, BC, V1Z 1N9
Telephone: 250 769-8803
www.mtboucherie.bc.ca

Quails' Gate Estate Winery
3303 Boucherie Road, Kelowna, BC, V1Z 2H3
Telephone: 250 769-4451
www.quailsgate.com

FROM SUMMERLAND TO PEACHLAND

"THERE IS A DANGER: YOU DON'T WANT TO BE TOO ELITIST AND MAKE ONLY WINES THAT MOST PEOPLE WOULDN'T ENJOY. BUT IT IS NICE TO OFFER THINGS THAT PEOPLE WILL NOT FIND AT MOST OTHER WINERIES AND THAT PEOPLE CAN APPRECIATE."

—TILMAN HAINLE
HAINLE ESTATE WINERY

The base camp for touring this group of wineries is the fragrant ornamental garden at the federal government's agricultural research station south of Summerland. High on a plateau, the station commands superb views of Okanagan Lake and, on the far side, the vineyards of the Naramata Bench. The back of the ornamental garden (expertly tended now by volunteers) looks onto the Trout Creek trestle of the former Kettle Valley Railway. Here, the rail bed, still with a few miles of steel rail, winds behind Giant's Head, the extinct volcano towering over Summerland. Beyond the trestle is a valley of orchards and vineyards, with two wineries — Scherzinger Vineyards and Thornhaven Estates — that may be perceived as being off the beaten track even if they are just 10 minutes from Highway 97. There was once bad blood between Scherzinger and Sumac Ridge, visibly located on the highway, and Sumac Ridge employees were forbidden to provide directions to Scherzinger. Now winery signs on the highway make it easy.

For more than 30 years, important technical work for the wine industry has been done at what many still call the Summerland research station, because the proper name is such a mouthful: Pacific Agri-Food Research Centre. It was established in 1914 and, for much of its history, focussed on the once-dominant tree fruit industry. Grape-breeding

OPPOSITE: **BEHIND GIANT'S HEAD AT SUMMERLAND, THE THORNHAVEN WINERY'S VINEYARD SLOPES STEEPLY TO THE SOUTH.**

trials began in the early 1970s to find winter-hardy varieties with which the fledgling wine industry could "stem the tide of imports by producing grapes with desirable table and wine characteristics," as a research station history puts it. The program produced at least one successful table grape, Sovereign Coronation, and one wine variety, Sovereign Opal, which was released for grower trials in 1976. It survives today on one Kelowna vineyard and, as a wine, only in the Calona Vineyards portfolio. The centre also imported varieties to the Okanagan from elsewhere, including, in the 1970s, hardy grape vines from the Soviet Union. Their unpronounceable names would have handicapped them in the market if winemaking trials, conducted at Brights, had succeeded. (There may still be hope for one of these varieties. The owners of the Mt. Boucherie winery at Westbank have planted Michurnitz, a grape with an exceptionally dark red juice.) With growers themselves taking the risk of planting classic European varieties in the Okanagan's vineyards, the research centre's work turned to more fundamental concerns, such as classifying soils. As well, the Vintners Quality Alliance tasting panel meets at the station monthly to pass judgment on most of the wines grown in British Columbia before they reach the market. There are no public tours or tastings at the station but the well-shaded ornamental garden is ideal for summertime picnics.

Trout Creek boils through its rugged course behind the garden before dropping toward the lake, tamed by a rock-walled channel that keeps the waters from meandering across this flat fan of land below the plateau. Near the mouth of Trout Creek and just beside the highway, the Adora Estate Winery opened in 2002, visible more because of its 10 acres (four hectares) of new vineyard than for its bland, metal-clad industrial building. The winery's three partners include two Vancouver businessmen: Reid Jenkins, a software entrepreneur who acquired an enthusiasm for wine in the mid-1990s when his firm designed the first Web sites for Sumac Ridge and Hawthorne Mountain Vineyards; and Kevin Golka, the owner of several car rental franchises. The third partner is the Okanagan's busiest winemaker, Eric von Krosigk. The partners chose this site, a former orchard, precisely because it is beside the same highway that delivers as many as 100,000 visitors to Sumac Ridge each year. Adora's plain vanilla building is the first step toward a more ambitious winery when the business justifies it. "I told them they wouldn't be making any money in the first 10 years," von Krosigk says of his partners.

Born in Vernon in 1962, von Krosigk trained at Geisenheim and apprenticed with a German producer of sparkling wine. He returned home proselytizing sparkling wine. "I've been preaching the culture of sparkling wines for a long time," he says. "It's fabulous with lunch. Even with breakfast. It's good basically before or after sleep. Bubbly is such a fun wine. Bubbly has other qualities than your average table wine. It tends to get people animated, awake and alive. That is often why it is served at the beginning of functions. Bubbly is really about life." He is or has been the winemaker at Summerhill, Hawthorne Mountain, Pinot Reach, Red Rooster, Hillside, Saturna Island Vineyards, Godfrey-Brownell, Victoria Estate Winery and Marley Farm (and the list is incomplete). Almost every one of these producers makes sparkling wine. Adora is no exception.

Adora will push the envelope with small lots of stylistically individual wines. For example, the winery's Pinot Blanc, richly textured with honeyed fruit but a dry finish, departs significantly from the crisp green-apple style of most Okanagan Pinot Blancs. "It's not what you would say is typical but I didn't want to be typical," von Krosigk says. The blended wines also are singular. Elements No. 8, a white wine blended with eight varieties, is built with layer upon layer of flavour. "Like a mood ring," the winemaker says. Maximus, the winery's blended red, incorporates five grapes. "That's the whole art of blending," von Krosigk says. "Putting all the spice points in and really creating something you have a vision for two, three, ten years down the road."

As if he hasn't enough on his plate, von Krosigk runs his own six-acre (2.4-hectare) vineyard and has registered the name Summerland Estate Winery. "I keep threatening to open it up, but I see so many people in trouble because they have

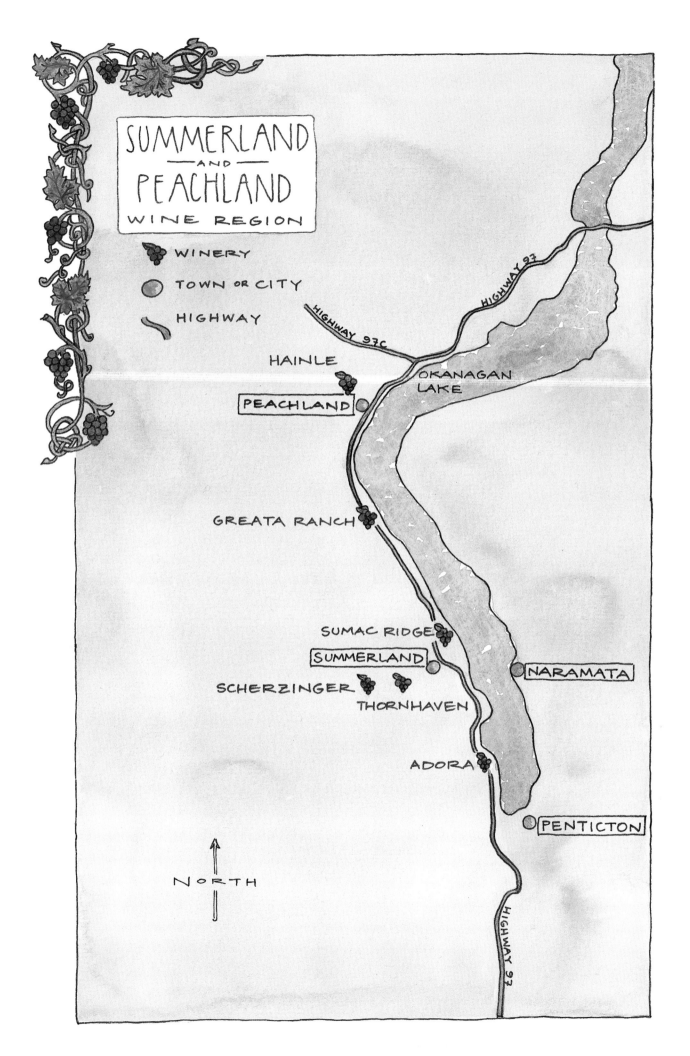

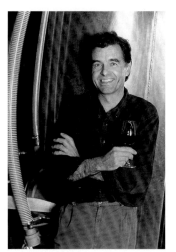

TILMAN HAINLE
HAINLE VINEYARDS

RON WATKINS AND EDGAR SCHERZINGER
SCHERZINGER VINEYARDS

HARRY MCWATTERS
SUMAC RIDGE ESTATE WINERY

ERIC VON KROSIGK
HILLSIDE ESTATE

CLOCKWISE FROM TOP: **SHAWNY, ALEX, PAMALA AND DENNIS FRASER**
THORNHAVEN ESTATES WINERY

overextended themselves," he says "As far as I'm concerned it is not a race. I'll open it up one of these years but I don't know when."

A few hundred yards along the highway from Adora are the signs directing visitors to both Scherzinger Vineyards and Thornhaven Estates, tucked on the south side of Giant's Head. For some years, Edgar Scherzinger's small vineyard just beside the Kettle Valley Railway was one of Sumac Ridge's sources of Gewürztraminer (almost always the big winery's top-selling varietal). McWatters was extremely annoyed when he was not informed in 1995 that this prized grower no longer intended to honour his contract to deliver grapes to Sumac Ridge. The staff at Sumac Ridge was discouraged from providing directions to Scherzinger Vineyards. The episode speaks to the quality of grapes that Scherzinger was growing on what once was a cherry orchard he bought in 1974. After losing money on cherries for several years, he switched to grapes in 1978, ultimately planting both Gewürztraminer and Pinot Noir on the south-facing slope and floor of a little valley. An accomplished wood carver from Germany's Black Forest, he built his workshop on the edge of the valley. In spite of being improbably small, this became both the winery and tasting room, with Scherzinger and his family offering visitors both wines and carvings. When he retired, the winery was acquired in May 2001 by Ron Watkins and his wife, Cher, a banker.

Ron Watkins was born in Winnipeg in 1950. He grew up in California (and remembers the wine there at the time as "pretty bad stuff"), returning to Canada in 1968. After working on Okanagan orchards, he turned to construction and specialized in energy-efficient buildings made from very tightly packed bales of straw. "You just stack them like bricks," he says. He has built two-storey homes with this technique, which he says is quick, strong and safe. When he gets around to rebuilding the tiny Scherzinger winery, now producing 2,400 cases a year, he intends to do so with bales of straw.

THE KETTLE VALLEY RAILWAY'S 1924 SHAY STEAM TRAIN GLIDES THROUGH VINEYARDS NEAR SUMMERLAND.

Watkins was pursuing his construction career in southeastern British Columbia and was seriously planning a vineyard there when Scherzinger, a long-time friend, offered to sell him the winery. Watkins was mentored by Scherzinger over three years before taking control of the winery and its seven acres (2.8 hectares) of vineyard, two-thirds of which is planted to Gewürztraminer and the remainder to Chardonnay and Pinot Noir. Like Scherzinger, Watkins makes Gewürztraminer in styles from dry to a rosé called Sweet Caroline, in which Pinot Noir is added to the blend. Five of the winery's nine wines are off-dry and as rich in flavours as the vineyard can deliver. "I'm a hanger," Watkins says of his propensity to leave the grapes maturing on the vines late into the year. "We go for the sugar content." Watkins also buys grapes from elsewhere in the Okanagan, including Merlot from the Naramata Bench and Riesling from the Golden Mile.

In the same month that Watkins took over Scherzinger, the Fraser family opened Thornhaven Estates Winery nearby. Dennis Fraser and his wife Pamala operated a 2000-acre (809-hectare) grain farm near Dawson Creek, B.C., until moving to Summerland. In 1994, they purchased an eight and a half-acre (3.4-hectare) apple orchard, replanting the steep slope (grades up to 25 percent) with Gewürztraminer, Chardonnay, Sauvignon Blanc, Pinot Noir and Pinot Meunier. Son Alex Fraser, born in 1975, had a developing career in northern British Columbia as a chef until his father recruited him to the winery. The first vintage was made in 1999. The coolly confident younger Fraser learned winemaking with consulting winemakers Ross Mirko and Christine Leroux. The winemaking style is clean and mainstream. Alex did once agree to

ONCE A MAJOR ORCHARD, GREATA RANCH NEAR PEACHLAND GROWS PREMIUM GRAPES FOR CEDARCREEK.

make some Pinot Noir in stainless steel when his father expressed a preference for non-oaked wine. The young winemaker also put some in barrels and the difference in quality was so evident that all Thornhaven Pinot Noir now sees barrels.

The 4,000-square-foot (371-square-metre) winery is a splendid Santa Fe–style building nestled harmoniously into a terrain of dry grass, pines and cactus plants. Indeed, the Frasers planned to call the winery Cactus Creek until they discovered the name was owned by someone who wanted $5,000 for it. They reached into their Scots heritage to come up with Thornhaven. The winery's patio, available for wine by the glass with delicatessen lunches or do-it-yourself picnics, looks west over the neat rows of vines. Because their previous expertise was growing wheat, the Frasers retained Valerie Tait, one of the Okanagan's busiest grape-growing consultants, to ensure that Alex Fraser has quality fruit to work with in the winery. "We don't go for volume," Fraser says. "We just go for nice, ripe fruit." In the 2000 vintage, when the vineyard's Chardonnay was producing the first time, the vines were thinned with such discipline that the 2,700 plants yielded just 2,200 bottles of finely concentrated wine. The wines are all from estate-grown fruit. Fraser expects the vineyard can produce enough fruit for 2,200 cases of wine, a little more than half the capacity of his cellars. Grapes from nearby vineyards and from elsewhere in the Okanagan will allow Thornhaven's production to increase as the market demands. In the winery's relaxed tasting room, Alex Fraser and his wife, Shawny, a banker, seem in no hurry to grow beyond what the family members can handle.

When Scherzinger's Watkins took over the winery, he shrewdly patched up relations immediately with Sumac Ridge's forceful founder, Harry McWatters. A gifted marketer, at the age of 23 McWatters began selling wine in 1968 when he joined the two-year-old Casabello winery in Penticton. Casabello succeeded in part because its baronial winery and sales room were right on the city's Main Street (the winery was dismantled in the 1990s after Vincor acquired the business). McWatters understood that location is crucial when he and then partner Lloyd Schmidt formed Sumac Ridge in 1980. They purchased the Sumac Ridge Golf and Country Club, with a nine-hole course at the edge of Summerland, right beside the highway. Fairways one and two were relocated to accommodate seven acres (2.8 hectares) of vines and the clubhouse did double duty as the winery and tasting room.

Initially, Sumac Ridge kept the golf club open, not just for cash flow but because its restaurant showcased the wines. The golf course was sold in 1990. When the golfers built a new clubhouse, their former clubhouse became the winery's Cellar Door Bistro, offering elegant wine-country dining. The formerly unpretentious clubhouse turned winery has transformed into a château on the hill beside the highway. The cellars have been expanded to accommodate 2,000 barrels and more racks for sparkling wines. A major extension in 2002 increased the size of the restaurant, more than doubled the sales room and added second-floor tasting rooms. The former compact tasting room had become so crowded that, in 2001, McWatters observed some visitors unable to enter and others abandoning wines they wished to buy because the cashier's line was too long.

While the original golf course vineyard still grows Riesling, Pinot Noir and Gewürztraminer, the winery's major source of grapes is the 115-acre (46.5-hectare) Black Sage vineyard south of Oliver. Acquired as raw land in 1992, it is planted entirely to vinifera grapes for premium wines such as the $50-a-bottle Pinnacle, the prestige red added to Sumac Ridge's line with the 1997 vintage. The winery's tour program offers a vast array of wines, including special tastings for its Connoisseur's Club. Membership is free but there is a modest fee for the tutored tastings.

Sumac Ridge is one of the wineries where von Krosigk, still a student, participated in making trial lots of bottle-fermented sparkling wine in the late 1980s. The winery released its flagship Steller's Jay Cuvée in 1989, tentatively making a few hundred cases. A decade later the winery was producing about 5,500 cases of this wine each year. Named for British Columbia's provincial bird, the wine is a

ICEWINE GRAPES AT HAINLE VINEYARDS, CANADA'S PIONEER ICEWINE PRODUCER.

blend of Pinot Blanc, Pinot Noir and Chardonnay. "We're the only producer of sparkling wine in the world with a shortage," McWatters says, alluding to the surplus of sparkling wines after millennium demand around the world fell short of expectations. Sometimes when he leads a winery tour at Sumac Ridge, McWatters indulges in an old French party trick of opening Champagne with a sharp tap of a sabre to snap the chilled neck from the bottle.

North of Summerland, there is little vineyard land until CedarCreek's Greata Ranch vineyard just north of Okanagan Provincial Park. The planned 2003 opening of the small Greata Ranch winery is in response to obvious demand. Since the Greata Ranch vines began producing, hundreds of visitors have called at the home of vineyard manager Merle Lawrence, seeking to taste and buy wine.

The 110-acre (44.5-hectare) property, sprawled on a plateau high above the lake, was once one of the most successful orchards in the Okanagan. It is named for George Greata, an immigrant from Britain in 1895 who planted apple trees here. The Long family, who bought it in 1910 and ran it until 1965, built their own packing house with a dock for shipping the orchard's fruit by boat. The severe winter of 1964–65 killed most of the fruit trees (and many of the Okanagan's vineyards) and the ranch, after several residential developments failed, fell into neglect until CedarCreek acquired it in 1994. CedarCreek owner Ross Fitzpatrick, whose father had managed other Okanagan packing houses, remembered buying fine peaches from Greata. He reasoned that it could produce equally fine grapes. Now, 32 acres (13 hectares) is planted with Gewürztraminer, Pinot Blanc, Merlot and several clones of Pinot Noir and Chardonnay. Plans for the remainder of the property, which includes a long stretch of private beach, are for an elegant housing development.

Unlike Greata Ranch, the vineyards above the lakefront community of Peachland are not visible from the highway but require a detour uphill, into a residential community. Trepannier Bench Road, at the northern edge of Peachland, once gave access to three wineries. Chateau Jonn de Trepannier, the Okanagan's first estate winery, was established at the end of the road in 1979. The winery went through several name changes and receiverships before closing, but the vineyard is still there. Nothing remains of A & H Vineyards, a small winery halfway up the road that opened in 1990 and closed in 1995. Hainle Vineyards, just up the road from the highway, remains open, serving lunches and dinners in its casually elegant Amphora Bistro, which opened in 1995 to showcase the often distinctively dry wines.

In 1972, the late Walter Hainle and his wife, Regina, who had emigrated several years earlier from Germany, bought an unserviced former homestead on the hillside overlooking Peachland. A textile salesman who had become an amateur winemaker in Canada, Hainle began developing a vineyard two years later. "It was still going to be just a hobby but my Dad was always hatching those big ideas," his son, Tilman, remembers. The "idea" sent

Tilman back to Germany to qualify as a winemaker. His family acquired additional vineyard in 1981, after he had returned, and opened the winery in 1988. It remained in family hands until 2002, when it was acquired by Walter Huber, another immigrant from Germany. A successful operator of fishing lodges in northern Ontario, Huber already owned another steep hillside property above Peachland for an intended winery. He planted grapes there after buying the Hainle winery.

Tilman Hainle, a distinctive stylist, remains the winemaker. Hainle is perhaps best known for making austere, concentrated and long-lived Riesling wines expressing the particular terroir of these Peachland vineyards. "The whole area we have planted around here is glacial till," he says. "The benchlands around here are sandy and rocky. In 1976 we drilled for water on the lower vineyard and went down 280 feet through sand and gravel before we found water." The vines struggle in the nutritiously lean soils. Hainle attributes the "steeliness" of his Riesling to this soil. By keeping the water and fertilizer to a minimum, Hainle forces the stressed vines to grow berries of concentrated flavour. In these vineyards, Hainle also grows Chardonnay, Traminer, Pinot Blanc and Pinot Noir.

"The Chardonnay, when we are left with any [bears raid the vineyard], has great quality and flavour," the winemaker says. "The nicest thing you can say about a Chardonnay is that it is Riesling-like. I look for some of that citrus quality in there and some of that austerity. If it is all flowery, bright and full for a Chardonnay, it does not ring true with me."

Beginning in 1993, the Hainle vineyards and the winery practises have been converted to organic techniques. Tilman Hainle believes that this has resulted in more powerfully flavoured wines. That bonus, however, is not the reason for his organic commitment. "To me, it's much more the ideal and the knowledge that you're trying to do the right thing which is the motivating factor." Hainle says. "The one consistent message that people come back to us with is that the flavour concentration and character in these wines is stronger than they normally encounter with other wines. When people come here, the main thing that we focus on is that the wines are specialized in style: dry, full-bodied, food-friendly…and only then we might mention organic. People should come to us for the style of wine, and organic should be the icing on the cake. But I am also glad when people come to us and say they want to support organic."

THE WINERIES

Adora Estate Winery
6807 Highway 97, Summerland, BC, V0H 1Z0
Telephone: 250 404-4200

**Dirty Laundry Vineyard
(formerly Scherzinger Vineyards)**
7311 Fiske Street, Summerland, BC, V0H 1Z0
Telephone: 250 494-8815 E-Mail: scherzi@telus.net
www.dirtylaundry.ca

Greata Ranch Vineyards
697 Highway 97S, Peachland, BC, V0H 1X9
Telephone: 250 767-2605

Hainle Vineyards
5355 Trepanier Bench Rd., Peachland, BC, V0H 1X2
Telephone: 250 767-2525
www.hainle.com

Sumac Ridge Estate Winery
17403 Highway 97, Summerland, BC, V0H 1Z0
Telephone: 250 494-0451
www.sumacridge.com

Thornhaven Estates Winery
6816 Andrew Avenue, Summerland, BC, V0H 1Z0
Telephone: 250 494-7778
www.thornhaven.com

PENTICTON AND THE NARAMATA BENCH

"OUR WINES REFLECT OUR TASTE IN THE WINES WE ENJOY DRINKING....
THE LUXURY OF BEING SMALL IS THAT YOU CAN AFFORD TO DO THAT."

—BOB FERGUSON
KETTLE VALLEY WINERY

The vineyards of the Naramata Bench, on the east side of Okanagan Lake, can be taken in at a glance from Agriculture Canada's Summerland research station, one and a half miles (2.5 kilometres) across the lake. The immense lake's capacity to store warmth and reflect light onto the vines makes this sandy loam one of the Okanagan's best vineyard locations. Unlike the vineyards near Oliver, where there is no tempering lake effect, grapes grown here have a long frost-free autumn in which to mature. Since 1990, when Lang Vineyards and Hillside Cellars (as it was then called) became the first wineries on the Bench, a dozen wineries have wedged onto the narrow stretch running from Penticton to Paradise Ranch and Okanagan Mountain Provincial Park. "This eastern side of the lake has a long day length," Australian-born winemaker Jeff Martin says, giving one reason for locating his new La Frenz winery on the Naramata Bench. "This is a better growing side than the west side. Not a lot of vineyards over there really have a track record for producing super-premium grapes." Ian Sutherland, Poplar Grove's irreverent owner, calls Summerland "the dark side."

A more leisurely view of the Bench is available from the abandoned Kettle Valley Railroad. Its roadbed, now a well-used hiking and cycling trail, defines the upper elevation of the vineyards. Several producers, including Lang, Hillside, Lake

OPPOSITE: **RED ROOSTER'S NARAMATA VINEYARD CATCHES THE LAST RAYS OF A LONG SUMMER'S DAY.**

WINERY BOTTLING LINES ARE MARVELS OF AUTOMATION.

Breeze, Red Rooster and La Frenz, have tasting rooms or dining patios looking out across the vineyards and lake, an aspect that Martin considers one of the best views in the world. The Bench is a wine touring destination, with resorts, bed and breakfasts, fine country dining and a restored heritage hotel in sleepy Naramata. (According to Rosemary Neering's *A Traveller's Guide to Historic British Columbia,* John Moore Robinson, the spiritualist developer who founded the village, intended to call it Brighton Beach until, during a séance, he heard the name Naramatah, referring to the wife of a Sioux chief. The name means "Smile of the Manitou.")

While Paul Gardner's Pentâge Winery, at the southern edge of Penticton's city limits, is not on the Bench, it is central to a pocket of steeply sloping vineyards with a comparable aspect toward the smaller Skaha Lake. These vineyards, including Gardner's undulating six-acre (2.4-hectare) Vista Ridge Vineyard, capture additional heat from the Skaha Bluffs, the gneissic cliffs that rise sharply above the valley not far behind the vineyards. The bluffs are renowned among rock climbers, many of whom, Gardner hopes, will slake their thirst at Pentâge.

Born in 1961 and a marine engineer who spent nearly 20 years on Pacific Coast towboats, Gardner decided to change careers in the 1990s. He and his partner, Julie Rennie, were enchanted by this rugged location with its shimmering lake view when they walked the property in 1996. Decades earlier, this had been part of Braeside Farms, one of the largest apricot farms in the world. By the time Gardner found it, it was the "brownest farm in the Okanagan," lying idle after an unsuccessful attempt to put housing on it. Gardner contoured the rugged property with a bulldozer and planted in 1997. The varieties chosen included two whites (Gewürztraminer and Sauvignon Blanc) and five reds (including Syrah), inspiring him to adapt the Latin for five to the winery name: Pentâge. For his first vintages, Gardner took over the four-car garage at his father's house after Norman Gardner, an aeronautical engineer, retired to a property just up the hill from Pentâge and obligingly planted a small Pinot Gris vineyard. Subsequently, winemaking moved into a smartly designed machine shed at Pentâge. This is an interim step. Paul Gardner has blasted a small canyon into a rock outcrop in the vineyard. The final winery will be set into this crevice, with an adjacent tasting room taking full advantage of the view.

When Gardner, an amateur winemaker, was ready in 1999 to make the first wines at Pentâge, he secured help from Sumac Ridge winemaker Mark Wendenburg, the owner of a nearby vineyard. Subsequently, Gardner has been mentored by consulting winemaker Ross Mirko. "I tend to like to create things," Gardner says. "The challenge is what motivates me." But he understands that the foremost challenge is getting it right in the vineyard first. "If the grapes are good, it does at times even seem easy."

Benchland Vineyards is within the northern border of Penticton's limits. The vineyards — nine

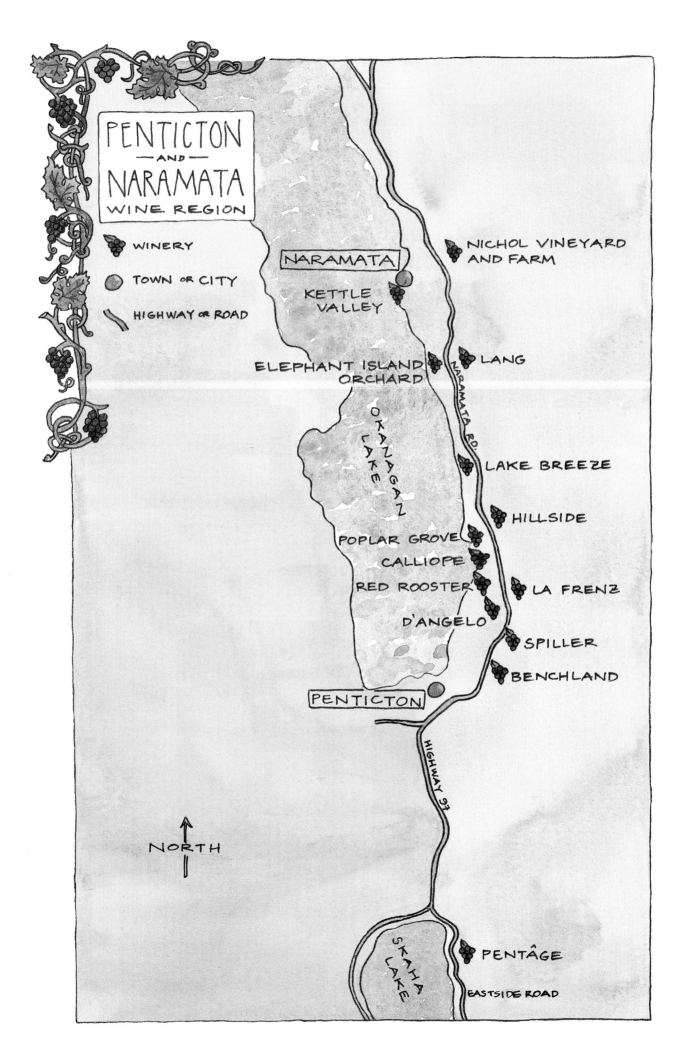

KLAUS STADLER
BENCHLAND VINEYARDS

TIM WATTS AND BOB FERGUSON
KETTLE VALLEY WINERY

KATHLEEN AND ALEX NICHOL
NICHOL VINEYARD & FARM WINERY

GITTA AND IAN SUTHERLAND
POPLAR GROVE WINERY

GÜNTHER LANG
LANG VINEYARDS

acres (3.6 hectares) owned by the winery and 10 acres (four hectares) on leased land next door — are on former orchard property, reasonably level and fertile. The winery was opened in late 1999 by Klaus Stadler, a restless brewmaster who was born in 1955 south of Munich and immigrated to Canada in 1997 for "a little change in life." Stadler, who had made spirits in Germany, also wanted to open a distillery. "I think products distilled from Okanagan fruit could be among the best in the world," he asserts. "And the orchardists are just crying for markets." Sadly, he shelved plans for grappa and eau-de-vie, discouraged by the brutal taxes on distilling. His winery is a low, earth-toned building that blends into the urban neighbourhood. The winemaking equipment is polished stainless steel because Stadler, fanatic about cleanliness, views barrels as dated technology, difficult to keep clean. When challenged on that view, Stadler snaps: "What model car do you drive? Why don't you drive a 1956?"

In the vineyards, planted during four years beginning in 1998, Stadler was the first in British Columbia to plant Zweigelt, a red variety developed and grown widely in Austria. He ordered Cabernet Sauvignon, Merlot and Pinot Noir as the main reds, along with three white varieties, Riesling, Pinot Blanc and Chardonnay. The supplier was unable to provide all the requested Pinot Noir and offered Zweigelt vines. Stadler had tasted many easy-drinking examples in Austria. "I was so happy to get Zweigelt," he says. On its own, the Benchland Zweigelt is a juicy, unoaked red with tart cherry flavours. It also is an element, along with Lemberger and Cabernet Sauvignon, in Mephisto, the winery's premium blended red, where Stadler uses chips to infuse an undertone of oak to the wine.

PERCHED HIGH ABOVE THE LAKE, THE KING FAMILY'S VINEYARDS ARE TYPICAL OF THE NARAMATA BENCH.

JEFF MARTIN
LA FRENZ

GARRON ELMES
LAKE BREEZE VINEYARDS

PRUDENCE AND BEAT MAHRER
RED ROOSTER WINERY

A little farther along the Naramata Road, Lynn and Keith Holman are opening the Spiller Estate Fruit Winery. The tasting room is in a heritage house once occupied by a Penticton pioneer named Albert Spiller, who owned a substantial ranch here as well as a sawmill that made ties for the Kettle Valley Railway. The intersection of Upper Bench Road and Naramata Road is known locally as Spiller's Corner and the Holmans searched no further for a winery name. They have a long tradition as orchardists, beginning with Lynn Holman's grandfather, Cedric Sworder, another of Penticton's pioneers. The Holmans, fruit growers for more than 25 years, grow primarily cherries and apples on their 100 acres (40 hectares). They opened the winery and the bed and breakfast for a practical reason. "It happened mainly because you can't make a living fruit farming," Lynn Holman says. Under the direction of veteran winemaker Ron Taylor, Spiller produces dry peach, pear and apple wines as well as dessert wines from apricot, peach, pear and cherry. The Holmans also leased a five-acre (two-hectare) vineyard nearby, enabling Taylor to make several blended wines and a Pinot Noir rosé. The Holmans plan a small vineyard on their own property as well.

The next winery along the Naramata Road is La Frenz, opened in 2002 by Jeff Martin and his wife, Niva. "It is good dirt," Martin says of his site. "And the beauty of it: the aesthetics are just incredible." Born in Australia in 1957, Martin was making premium wines for McWilliams when the Quails' Gate winery in the Okanagan recruited him in 1994. After making five vintages that established Quails' Gate's

OPPOSITE: **FRITZ HOLLENBACH'S VINEYARD SOUTH OF PENTICTON LOOKS ONTO SKAHA LAKE.**

reputation, Martin returned to his native land late in 1998 to open a winery. Within six months, he fled Australia and its surfeit of wineries, returning to British Columbia. "In my 20-some years in the industry, it would have been the worst time to start a winery back there," he says. Martin's La Frenz winery — the name was his grandfather's surname — husbanded its finances by making its first three vintages with purchased grapes at the nearby Poplar Grove winery. "We've put all our life savings in this," he says. "We don't have partners. My resources have to go as far as possible." In 2002, Martin replaced an apple orchard with the six-acre (2.4-hectare) La Frenz vineyard, planting Shiraz, Merlot, Viognier and, for some innovative late harvest wine, the seldom-planted Schönburger.

Poplar Grove Winery, farther along the Naramata Road and at the end of Poplar Grove Road, inhabits another former apple orchard on a warm plateau 200 feet (61 metres) above the lake. Ian and Gitta Sutherland, the owners, were novices —"completely unprepared" in Ian's words — when the vines arrived in the spring of 1993. A Montreal native, he is a pipefitter by trade and his Danish wife is a nurse. When Sutherland turned on the irrigation system after the vineyard was planted, nothing happened. He discovered that he had installed 2,500 water emitters backwards. But the Sutherlands have seldom put a foot wrong since. Poplar Grove's first releases, a 1995 Cabernet Franc and a 1995 Merlot, won gold and silver medals at a subsequent Okanagan Wine Festival. Sutherland honed his natural winemaking talent by working the crush at least five times at wineries in Australia and New Zealand. When he was in Australia in 2001, he became enamoured with a local cheesemaker's products. "How hard can that be?" he asked himself. By May 2002, he had added a lean-to on the winery, enlisted a cheese-making partner and launched Poplar Grove Cheese with Naramata Blue and Okanagan Valley Cream (Camembert-style) cheeses. "What fuels this has been the same as what fuelled our winemaking — raw enthusiasm," he admits.

Poplar Grove grows ripe red wines that, Sutherland maintains, reflect the particular terroir of its nine acres (3.6 hectares) of vineyard. "The soil is the kind one would find in Pomerol or St-Emilion," he says. "It's heavier soil, so it produces Merlots and Cabernet Francs with body to them. And sometimes I have noticed in our wines that there are notes of sage in the reds. When we first cleared the land, the sage was predominantly on the bank. Now, it has migrated into the vineyard and is in the rows with the grape plants — and we leave it there for a reason. It adds an element to the local terroir that you aren't going to find in reds anywhere else. It's a fabulous, savoury, earthy flavour." The vineyard also includes Pinot Gris and a modest planting of Viognier, with purchased grapes providing the winery's Chardonnay. Sutherland produces between 1,500 and 2,000 cases of wine, priced to reflect that Poplar Grove now releases its reds only in their third year. Even though the wines sell quickly, Sutherland is not ambitious to be larger. "At the end of the day, one of the hardest things to recognize is when you arrive at where you want to be," he says. "Empire-building is not in the cards with us."

The accommodating Sutherland shares his compact winery with other vintners from time to time. La Frenz was based at Poplar Grove for several years until its own winery opened in 2002. Sutherland then took in Calliope Vintners, a winery launched in the 1999 vintage by consulting winemakers Ross and Cherie Mirko, in partnership with viticulturist Valerie Tait and her spouse, Garth Purdy, a Kelowna businessman. The label has followed Ross Mirko's consulting assignments as he has worked for successive wineries, including Thornhaven and Blasted Church. This frugal arrangement saves Calliope, which made only 900 cases of wine in 2001, from tying up resources prematurely in a winery of its own. Born in Vancouver in 1960, Mirko, a psychology graduate, began his career in wine in 1988 in the laboratory of Andrés in Port Moody. In 1994 he earned a postgraduate enology diploma at Lincoln University, New Zealand's premier wine school, where Cherie, a native of New Zealand, was a classmate. On their return to Canada, they worked at several Okanagan wineries until becoming consultants and launching Calliope.

LAKE BREEZE WINE-MAKER GARRON ELMES CHECKS GRAPES JUST PRIOR TO CRUSH.

MOST BRITISH COLUMBIA VINEYARDS HAVE BEEN STARTED OR REPLANTED SINCE 1985.

Calliope — pronounced ka-LIE-oh-pee — is the name of a hummingbird indigenous to the Okanagan. When the winery sources grapes, it is also as free as a bird. The strategy is to buy premium grapes from growers throughout the Okanagan, relying on the detailed knowledge of Valerie Tait, who has worked as a consultant in the valley for a dozen years. Calliope's initial vintages have been made with fruit from a vineyard near Osoyoos Lake. The Calliope wines are made primarily with Bordeaux varieties. The reds are Cabernet Sauvignon and Merlot; the whites are Sémillon and Sauvignon Blanc, including blended wines. "We'd like to stick with these four varieties and not be the Baskin-Robbins of wine," Mirko says.

Not far from both La Frenz and Poplar Grove but closer to the lake, Ontario winemaker Salvatore D'Angelo will have converted a 14-acre (5.7-hectare) apple orchard to grapes by 2003. (The trees yielded 60 cords of wood to be used for barbecues at the planned winery guest house.) The son of immigrants who came to Canada in 1955 from Italy, D'Angelo planted his first vineyard near Windsor in 1979. He subsequently opened a winery there that now produces about 4,000 cases a year. Because he finds the dry Okanagan climate healthier than the Windsor area's humid summers, D'Angelo began coming west in 1985. At times, his rental car clocked as much as 1,900 miles (3,000 kilometres) in a week as he explored the wine country. "I should have bought in 1989 when everything was for sale," he reflects now. One evening early in the 1990s, at dinner with Vera and Bohumir Klokocka, who then owned Hillside, D'Angelo complained that he could find no good red wine in British Columbia as he produced a bottle of his gold medal–winning 1991 Cabernet Sauvignon from Ontario. His hosts

THE SILENCE OF WINTER LIES ACROSS DORMANT VINEYARDS ON THE NARAMATA BENCH.

responded with a tasty carafe of their Cabernet Sauvignon; Hillside in 1992 may have been the first Okanagan producer of this variety. "That's when it finally got into my head that you *can* make good reds in the Okanagan," he recalls. He has planted only reds: Cabernet Sauvignon, Cabernet Franc, Merlot and Pinot Noir. D'Angelo, 50, continues to operate his Ontario winery while he looks for more vineyard land in the Okanagan. "I want to experiment with Shiraz and some Italian reds," he says. His son, Christopher, 21, who has studied viticulture and winemaking at Okanagan University College, will look after D'Angelo Vineyards on the Naramata Bench.

Czech-born Vera Klokocka, whose wine inspired D'Angelo, sold Hillside after her husband's death in 1995. The winery has since expanded from Vera's humble farmhouse to an imposing 15,000-square-foot (1,394-square-metre) structure beside the Naramata Road. The location, which includes a restaurant, attracts more than 20,000 visitors each year, including cyclists from the Kettle Valley Trail whose purchases the winery will deliver directly to their hotels. That shows the business savvy of general manager Ken Lauzon, a Windsor native who came to the winery in 1999 from the restaurant and hospitality industry, latterly managing a private club in Calgary. "I know enough about what I need to know but I am not a winemaker," Lauzon says. "That doesn't even appeal to me. The end result is the most appealing part." With veteran winemaking consultant Eric von Krosigk in the Hillside cellar, Lauzon is focussed on selling the wines, sometimes quite creatively. In 2002, when Penticton hosted an Elvis Presley festival — the first of many, Lauzon hopes — Hillside won the contract to make the official wine. That was how Graceland Gamay, Hound Dog Chardonnay and Blue Suede Blush joined the Hillside line of wines.

By tradition, Hillside has specialized in whites, with the signature wine being the crisply fragrant Muscat Ottonel, the major planting in the adjacent three-acre (1.2-hectare) vineyard. Red wine production will rise as grapes come from Hillside's 12-acre (4.9-hectare) Hidden Valley Vineyard, south of the

THERE IS ALWAYS ANOTHER VINTAGE WORTH WAITING FOR.

winery, which was planted in 2002 to Syrah and Merlot, along with Gewürztraminer. In addition, Hillside also owns what it calls the Blackhawk Vineyard, a 60-acre (24.3-hectare) parcel south of Paul Gardner's Pentâge Winery. Also part of the former Braeside apricot farm, this is being planted over a period of four years.

The Lake Breeze winery has evolved grandly since 1994 when the first proprietor, former South African businessman Paul Moser, purchased 13 acres (5.3 hectares) of vineyard and a farmhouse on a plateau overlooking the lake. Within two years, Moser had recreated a neat but tiny white-stucco Cape-style winery on the property, hired a young South African winemaker named Garron Elmes and even planted British Columbia's first small plot, half an acre (.2 hectare), of Pinotage, South Africa's own red variety. Production at Lake Breeze ultimately outstripped the winery's capacity. In the spring of 2002, two business couples from Alberta took over Lake Breeze, immediately quadrupling the size of the winery and popular restaurant patio while still retaining the photogenic Cape architecture.

The Lake Breeze site was formerly part of the larger vineyard, best known for white varieties, which sweeps down a long, southwest-facing slope below Naramata Road. "We do whites a lot better up here than we do reds," Elmes maintains over tastings of Pinot Blanc (the flagship white from vines planted in 1984), Pinot Gris, Gewürztraminer, Chardonnay and Sémillon, all made in a fruit-forward style. He hopes to add Sauvignon Blanc while reducing the red varieties in the vineyard to Pinot Noir, Merlot and, of course, the exclusive Pinotage, first released with the 1998 vintage. "I go for a light to medium-bodied style," Elmes says of the lively, ruby-coloured and berry-flavoured Pinotage.

Nestled against a five-acre (two-hectare) orchard a short drive north of Lake Breeze is the Elephant Island Orchard Winery, with novel wines and a shaded picnic patio. Elephant Island's fruit wines — it makes only fruit wines — are memorably good. "We are really after the same thing as quality grape wineries are after," says Del Halladay who, with his wife, Miranda, opened the winery in 2001. "We use the best fruit we can." The wines deliver the flavours and aromas of the fresh fruit because water is not added to any fruit except for the acidic black currant. With moderate alcohols, often 10 to 11 percent, the wines, mostly dry, are pleasant both as apéritifs and with food. "A lot of people have a misconception that you cannot produce dry fruit wines," Halladay says. "Some of our dry wines are our most popular."

Both the wines and the winery name were inspired by Miranda's grandfather, Paul Wisnicki, an engineer, and his wife, Catherine, an architect. About the time the Hallidays were born (Del in Victoria in 1972 and Miranda in Powell River in 1973), Catherine Wisnicki bought this Naramata property as an investment. When she designed the house, her husband scoffed that her design was "all for the eye." Something of a punster, she christened the property Elephant Eyeland — which quickly became Island. Before his death in 1992, Paul Wisnicki, an amateur maker of fruit wines, had considered a distillery. Del Halladay, a marketing graduate from a Maryland university (he attended on a lacrosse scholarship and now plays the game professionally), tinkered with both his grandfather's business plan and fruit wine recipes to launch the winery. In what he describes as his best business decision, he recruited Christine Leroux, a French-trained winemaker, to provide "quality winemaking."

The fruits — pears, apples, cherries, apricots, crab apples and black currants, carefully chosen for their winemaking qualities — are all grow in British Columbia, primarily in the Okanagan. Halliday did trials with 30 varieties for the apple wine before settling on a blend of Golden Delicious, Granny Smith and crab apples. The pear wine employs Bartletts, with a little quince for more body. The Stella cherry is preferred for its high sugar and deep colour, both for dry and port-style wines. The Goldrich apricot has fallen from favour in the fresh market because the skin is acidic, but this makes it ideal for Elephant Island's well-balanced dessert wine. The bronze-pink crab apple wine, with a zesty finish like a Mosel Riesling, is made from an unnamed and rare pink-skinned variety. This is one of Elephant Island's most sought-after wines. Unfortunately, Halliday cannot find an abundant source of this crab apple. Meanwhile, he is thinking of adding rhubarb, blueberries and blackberries to Elephant Island's array of wines.

Lang Vineyards occupies perhaps the highest perch on the Bench, with the winery and tasting room overlooking the Riesling vines whose rows fall away toward the distant lake. Günther Lang and his wife, Kristina, purchased the nine and a half-acre (3.8-hectare) vineyard, then planted to hybrid varieties, when they immigrated to Canada in 1980 from Stuttgart. "I bought the vineyard because I fell in love with the place," Lang, formerly a manager with Mercedes Benz, says. By the middle of the decade, he was among the small band who badgered the provincial government until the farm winery license was created in 1989. The tiny but superbly equipped winery Lang opened the follow-

BIN OF GRAPES ABOUT TO BE DUMPED INTO WINERY'S CRUSHER-DESTEMMER.

WINERY CRUSHERS SEPARATE LEAVES AND STEMS FROM GRAPES BEFORE WINE IS MADE.

ing year has grown well beyond the farm winery concept. Lang hired a full-time winemaker in 1996 and the winery today makes about 8,000 cases of wine a year.

"We have a good location, no question," Lang says. It is cool enough to grow Gewürztraminer that retains refreshing acidity. With good air drainage down the slope, the vineyard remains free from frost well into the autumn, enabling Lang to ripen the Riesling for fine late harvest wines. The vineyard gets enough sun to ripen Pinot Meunier and Viognier, first planted on the Bench by Lang. However, the winery has enjoyed unique success with icewines and with dessert wines flavoured with maple syrup, an iconic Canadian product if ever there was one. "That was my own idea," says Lang, who started "as a very conservative thinker, making only varietals."

The winery's Original Canadian Maple red starts with a dry red blend of Merlot, Pinot Noir and Pinot Meunier while the white Canadian Maple's base is Pinot Auxerrois. Both are fermented to about 12 percent and finished as conventional wines. Then the wines are diluted with maple syrup (the volume of syrup in each wine is Lang's secret). The resulting wines are fruity and moderately sweet — but less than half as sweet as icewine and considerably less expensive. The wines have sold well in Asia and in duty-free shops. This success inspired the creative Lang to craft a sparkling wine, Canadian Maple Brut, in which maple syrup is used for added sweetness; the wine scored top acclaim in a national wine competition in 2001. Never content to rest on his laurels, Lang and his winemaker since have made a premium dry white called Grand Pinot because it employs grapes from the Pinot family. "And I've played with the idea of a grape liqueur," Lang says.

The fire hall on Naramata Road landmarks two wineries, Red Rooster and Kettle Valley, both located along Debeck Road, toward the lake. The road is named for the family who developed orchards here after World War I. In 1990, when they emigrated from Switzerland where they had operated fitness parlours, Beat and Prudence Mahrer bought one

RED ROOSTER'S PRUDENCE MAHRER TRIMS GRASS BETWEEN VINE ROWS.

OPPOSITE: **RELIC OF THE VINEYARDS IS NOW A LANDMARK AT RED ROOSTER.**

of the Debeck orchards, replacing the apple and apricot trees with grape vines. "Our neighbours thought we were crazy," Mahrer laughs now, relaxing in his gazebo in front of Red Rooster Winery with its breathtaking view of the lake. The compact winery farms 30 acres (12 hectares) and makes 10,000 cases a year. The hardworking Mahrers, both licensed pilots, still find time to explore British Columbia in their float- and ski-equipped Piper Super Cub aircraft. The winery's name, along with the chicken pen beside the driveway, arises from the affection developed for chickens when they kept several as pets in Switzerland.

"On the Naramata Bench, you are on the right side of the valley," Mahrer says of his southwest-sloping vineyard. "By growing on the slope, we are catching 60 percent more sunlight than on level

THE WINERY IS AN EXTENSION OF THE FARM.

OPPOSITE: **BIRD NETTING PROTECTS KETTLE VALLEY'S OLD MAIN VINEYARD AT NARAMATA.**

ground. We didn't know this when we started. We talked to everyone who knew something about grapes." The Red Rooster wines, in Mahrer's view, capture the Naramata terroir: crisp Chardonnay; Pinot Gris wines that are lively fruit bombs; Gewürztraminer with lichee flavours; and Merlot with bright berry notes. "It is the difference between eating an apricot from the tree and apricot jam," Mahrer says. "We are not making big, bold and fat wines."

In contrast, Bob Ferguson and Tim Watts at the Kettle Valley Winery at the end of Debeck Road make some of the boldest and longest-lived wines on the Naramata Bench. "Our wines certainly reflect our taste in the wines that we enjoy drinking," Ferguson says. "Ours have always been big, robust and full-bodied. We certainly have tried to make wines that are fairly intense." Their 1992 Pinot Noir, the debut vintage, remained deliciously alive a decade later. In 2001, they averaged a spare two and a quarter tons (two tonnes) of grapes from each acre (.4 hectare) of vineyard. "We're not farming to make money," Ferguson says. "We're farming to grow grapes to make wine — and we'll make our money from the wine." One-time amateur winemakers who are brothers-in-law, these are shrewd partners. Ferguson, a lean and lanky Scot, is a chartered accountant; the broad-framed Watts is a geologist. They opened their winery in 1996 in Ferguson's three-car garage, retaining that as the tasting room when, five years later and at a production of 4,000 cases, they moved winemaking to a plain, purpose-built structure nearby. The partners are not aggressively seeking to expand. "We want to be hands-on; we don't want to be here as the management team," Ferguson says. They would be tempted if a single large property on the Bench were available. Some of their vineyards are as small as an acre and are not always conveniently near each other. "We have been driving our tractor on the roads so much that we had to put heavy road tires on the front wheels," Watts says.

Nearly all of Kettle Valley's grapes come from Naramata Bench vineyards where they own or manage more than 20 acres (eight hectares). Significantly, nearly all those vineyards are downhill from the Naramata road, often close to the lake. The picturesque Old Main Vineyard, a stone's throw south of Naramata, perches on a west-facing bench, 100 feet (30 metres) above the lake. It was planted in 1991, primarily in major red varietals — Cabernet Sauvignon, Cabernet Franc, and Merlot — that are blended to make Old Main Red, one of Kettle Valley's signature reds. "The moderating influence of the lake makes a huge difference," Ferguson says. "We are able to let the fruit hang much longer in the fall." With little concern about frost, Kettle Valley harvests as late as mid-November, bringing in ripe, flavour-packed grapes.

The four and a half-acre (1.8-hectare) Nichol Vineyard, just north of Naramata, might well be one of the warmest sites on the Bench. Co-owner Alex

Nichol noted in a winery newsletter in 2001 that, in recording temperatures a decade earlier when he was first planting vines, he measured more days of heat than are reported in either Bordeaux or the Rhone. This accounts for the almost perpetual windburn he has sported since resigning as a double bass player with the Vancouver Symphony Orchestra to grow grapes. He and Kathleen, a librarian, bought this former pear orchard in 1988 and, below a heat-reflecting 300-foot (91-metre) granite cliff, planted the Okanagan's first Syrah. Subsequently, they added Cabernet Franc, Pinot Noir and Pinot Gris, along with a few rows of St. Laurent, an Austrian red variety orphaned when the provincial government ended its grape-growing trials and gave the vines to Nichol. From these and from purchased grapes, Nichol, a reliable and original winemaking stylist, now averages 1,200 cases of wine a year.

"I wanted to produce a big monster Syrah," Nichol recalls. "But that's not what this site gives me." The style he has achieved is much more like a Rhone red, somewhat lighter than he had in mind but, in the view of his peers, more elegant. On the other hand, his Pinot Gris can be what he calls "fat and pudgy" in some vintages, with alcohol exceeding 14 percent. It all reflects the site where Nichol discovered that "a macho vineyard" with heat spiking as high as 113°F (45°C) is not ideal. "We couldn't stand it and the plants couldn't either," says Kathleen. Most vines ripen grapes best at temperatures from about 77 to 84°F (25 to 29°C), above which the plants begin shutting down to conserve moisture, remaining in "hibernation" until temperatures drop. "What the plant is doing is protecting itself," Nichol reasons. Some varieties, such as Pinot Gris, tolerate more heat that either Cabernet Franc or Syrah. Nichol hit on the solution in 1999, installing overhead sprinklers and setting them to come on automatically at 84°F (29° C). The light shower only lasts five minutes but that is enough to cool the vines significantly, enabling them to keep working. "The plants look perkier, especially the Syrah," Nichol has found. This cooling technique is also used elsewhere in the Okanagan to moderate the heat of August.

About 6 miles (9 kilometres) north of Naramata, the road forks. The right fork heads to Okanagan Mountain Provincial Park and the left fork dead-ends at a breathtaking vineyard, the most northern of the Bench's vineyards. This 646-acre (261-hectare) property has been known as Paradise Ranch since it was homesteaded in 1904 as a cattle ranch. Vineyard development began in 1981. There now are about 100 acres (40 hectares) of vines, growing on a series of steep-sided benches that form a southwestern-facing amphitheatre stepping down toward a long beach. This vineyard is as precarious as it is beautiful; several tractors have tumbled over the edges of the benches, fortunately without serious injury to the operators. Because Paradise Ranch sits tightly against mountain wilderness, black bears occasionally forage among the vines, with a particular taste for ripe Merlot. A moonlight visit by one bear several years ago inspired the etching on all bottles of icewine produced by Paradise Ranch Vineyards.

There is no winery at the Ranch and there is a locked gate at the entrance. The property was purchased in September, 2002, by Mission Hill Family Estate, which had been buying grapes from the vineyard for 10 years. John Simes, the winemaker at Mission Hill, plans to expand the vineyard slightly, concentrating on growing white varieties and Pinot Noir.

Paradise Ranch, licensed as a winery since 1998, makes only icewine and late-harvest wine. Without a winery of its own, Paradise Ranch has hired existing wineries to produce its wines, which were made from grapes grown on the Ranch prior to 2002. While Mission Hill has purchased the property, the Paradise Ranch label survives. James Stewart, the Vancouver lawyer who owns Paradise Ranch Vineyards, intends to buy grapes and establish a winery elsewhere in the Okanagan.

THE WINERIES

** D'Angelo Vineyards
947 Lochore Road, Penticton, BC, V2A 8V1
Telephone: 250 493-1364

Elephant Island Orchard Wines
2730 Aikens Loop, Naramata, BC, V0H 1N0
Telephone: 250 496-5522
www.elephantislandwine.com

Hillside Estate
1350 Naramata Road, Penticton, BC, V2A 8T6
Telephone: 250 493-6274
www.hillsideestate.com

Kettle Valley Winery
2988 Hayman Road, Naramata, BC, V0H 1N0
Telephone: 250 496-5898
E-Mail: kettlevalleywinery@telus.net

La Frenz
740 Naramata Road, Penticton, BC, V2A 8T5
Telephone: 250 492-6690
www.lafrenzwinery.bc.ca

Lake Breeze Vineyards
930 Sammet Road, Naramata, BC, V0H 1N0
Telephone: 250 496-5659
E-Mail: lakebreeze@telus.net

Lang Vineyards
2493 Gammon Road, Naramata, BC, V0H 1N0
Telephone: 250 496-5987
www.langvineyards.com

Nichol Vineyard & Farm Winery
1285 Smethurst Road, Naramata, BC, V0H 1N0
Telephone: 250 496-5962

Paradise Ranch Vineyards
No winery.

Pentâge Winery
4400 Lakeside Road, Penticton, BC, V2A 8W3
Telephone: 250 493-4008

Poplar Grove Winery
1060 Poplar Grove Road, Penticton, BC, V2A 8T6
Telephone: 250 492-2352
www.poplargrove.ca

Red Rooster Winery
891 Naramata Road, Penticton, BC, V2A 8T5
Telephone: 250 492-2424
WWW.REDROOSTERWINERY.COM

Spiller Estate Fruit Winery
475 Upper Bench Road N., Penticton, BC, V2A 8T4
Telephone: 250 490-4162.

Stonehill Estate Winery
(formerly Benchland Vineyards)
170 Upper Bench Road S., Penticton, BC, V2A 8T1
Telephone: 250 770-1733
E-Mail: benchland@shaw.ca

** UNDER DEVELOPMENT

VINEYARDS OF OKANAGAN FALLS

"YOU HAVE TO HELP ONE ANOTHER OUT. THERE ARE NO SECRETS. I EVEN OFFER HELP TO AMATEUR WINEMAKERS. IT IS EVERYONE'S RIGHT TO PRODUCE WINE."
—FRANK SUPERNAK
BLASTED CHURCH VINEYARDS.

The rugged profile of McIntyre Bluff, rising 919 feet (280 metres) from the valley floor, pinches the Okanagan Valley about six miles (10 kilometres) north of Oliver. Named for Peter McIntyre, an early settler, the bluff is historically significant to the valley's First Nations people. It is said to have been the site of a decisive battle between the Okanagan and the invading Shuswap, in which only one Shuswap warrior survived.

To the wine growers of the Okanagan, McIntyre Bluff has another significance. The vineyards south of this great cliff and as far as the international border, grow in the hot Canadian extension of the Sonoran Desert. To the north, the vineyards between Vaseux and Skaha lakes have growing seasons that, on average, are several degrees cooler.

It is the difference between the sun-baked south of France and cooler Burgundy, according to veteran grape grower Ian Mavety, the owner with his family of Blue Mountain Vineyard and Cellars. That winery's vineyards are planted to the great Burgundy and Alsace grapes: Pinot Noir, Chardonnay, Gamay, Pinot Gris and Pinot Blanc.

It is anything but simple to pigeonhole the vineyards and wineries of the Okanagan Falls region. Blue Mountain succeeds with powerful Pinot Noir, elegant Chardonnay and subtle Pinot Gris. Wild Goose's rock-strewn vineyard produces

OPPOSITE: **UNDULATING ROWS OF VINES AT BLUE MOUNTAIN VINEYARD, WITH VASEUX LAKE AND MCINTYRE BLUFF IN THE DISTANCE.**

NEW LIFE EXPLODES FROM THE VINE BUDS EACH SPRING.

Riesling and Gewürztraminer, while its neighbour, Stag's Hollow, grows big Merlot. Hawthorne Mountain Vineyards, with a cool, northern-facing vineyard high above the valley, is renowned for its Gewürztraminer, while Blasted Church on the east side of Skaha Lake is still resolving an undisciplined melange of varieties. Almost every variety grown in the Okanagan can be found in one or another of the Okanagan Falls vineyards. There are two explanations for this. Some growers are still working out what varieties are best suited to this area; but also the terrain and the lakes create substantial differences in the growing conditions among the vineyards. However individual the wineries are, all share the area's picture postcard vistas and herds of both deer and handsome California bighorn sheep.

"You can't take Okanagan Falls and call it a viticultural region," says Ian Mavety, who has been growing grapes here since 1972. He once had Riesling and Gewürztraminer in his vineyard. He replaced them after being disappointed with trial batches of wine he made. "The wines were good as varietals but the potential was not there." The frugal Mavety sold those experimental wines in bulk to another winery. To his surprise, the buyer entered the wines in competition and won gold medals. Mavety, who can be hard-headed, stuck to his convictions about what is best on his site.

"It's very different on this bench than elsewhere," he says. He suggests that Oliver Ranch Road, which winds uphill from the highway and gives access to several wineries, divides the main soils of the area. Wild Goose and Stag's Hollow are on the gravelly soils west of the road to the Okanagan River. Blue Mountain and Eaglebluff, east of the road, are on sandy loam. The difference calls for different vines, with the result that, as elsewhere in the Okanagan, an abundance of varieties are grown. One grower is even experimenting with Zinfandel. "Why are we trying to be a supermarket?" Mavety grouses.

Lean and wind-burnt, the laconic Mavety approaches farming as a vocation. Upon graduating in agriculture from the University of British Columbia in 1971, he bought a rundown orchard south of Okanagan Falls with his wife Jane and planted the hybrid grape varieties then sought by the wineries. They began replacing these with vinifera in 1985, accelerating the conversion of the 65-acre (26.3-hectare) vineyard four years later in order to launch their winery. Careful and methodical, they sought pointers from vineyards in Europe and California while beginning winemaking trials on their own. They selected Burgundy and Alsace varieties because they concluded these would mature reliably in the Okanagan's intense but short season. "We could grow Merlot and make an interesting wine," Mavety says, referring to the Bordeaux variety that requires a slightly longer season. "But it would probably vary substantially from year to year. It would be marginal."

For Blue Mountain's 1991 vintage, the first commercial release when the winery opened in 1992, the winery retained a consultant, Raphael Brisbois, a French-trained winemaker then making sparkling wine at a top California winery. As a result, Blue

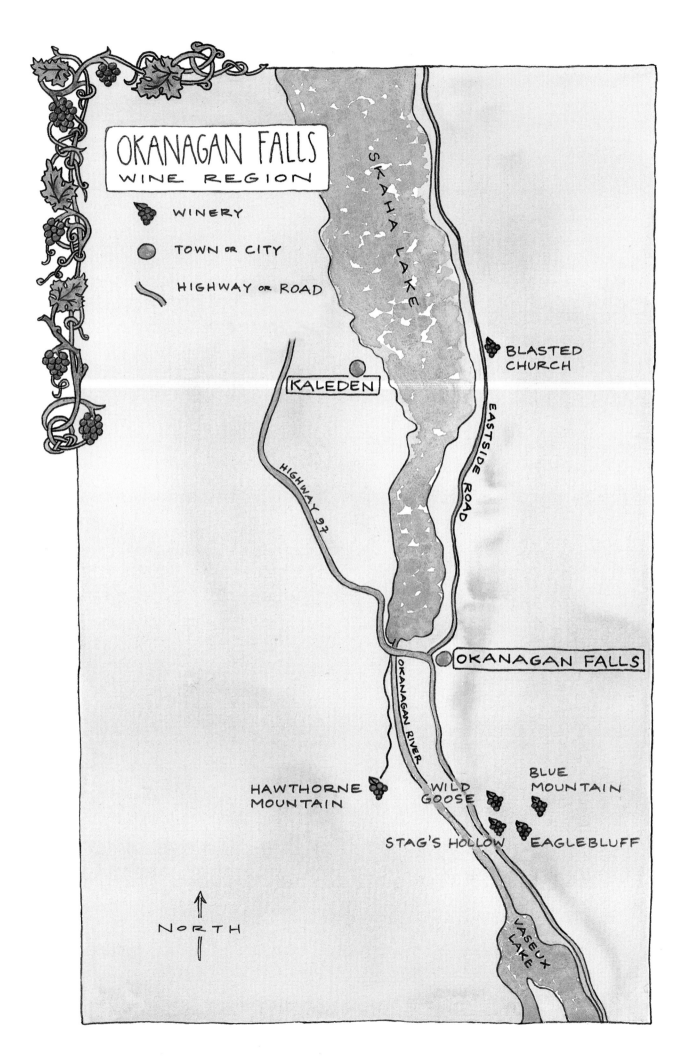

Mountain's wines, both still and sparkling, showed professional refinement. The table wines, especially those from Pinot Blanc and Pinot Gris, departed from the general Okanagan style at the time. "The winemakers in the valley then were making them sweet," Ian says. "We don't drink sweet wines. When Jane and I went to Europe, we realized that we didn't have to make sweet wines." The relationship with Brisbois continues but to a lesser degree, with winemaking in the hands of Ian Mavety and son Matt, whose agriculture degree includes a winemaking diploma from Lincoln College in New Zealand.

The wines continue to be refined. Contrary to the current fashion, Blue Mountain's wines are being made at slightly lower alcohol levels without sacrificing flavour. "It is a matter of cultural practices," Ian Mavety explains. Organic methods have been introduced to the vineyard. More recently, the Mavetys have begun to test what is called biodynamic farming, a term that encompasses pre-industrial methods applied in some European vineyards (such as burying stag bladders in compost or observing phases of the moon when carrying out vineyard or winery activities). "It's not very exciting stuff," Ian says. "This is peasant agriculture." The point, in addition to growing fine wine, is to farm responsibly. "The best expression of your site will be the least amount of intervention," he believes. "We both think we can rely less on chemical inputs."

The Kruger family's Wild Goose grows its fruit in soil dramatically different from that of Blue Mountain even though the wineries are only a 10-minute drive apart. One visitor to Wild Goose, on seeing the 10-acre (four-hectare) vineyard, asked if the stones had been placed there deliberately as heat sinks. "It is extremely hard to cultivate," says Adolf Kruger who, with his sons Roland and Hagen, first planted here in 1984. Holes for vineyard posts and for the vines could only be made with a high-pressure water gun. Hagen, who has succeeded his father as winemaker, was so physically exhausted after a weekend of planting that he had difficulty holding his cereal spoon on Monday

CLEANING GRAPE BINS AVOIDS PROBLEMS WITH FRUIT FLIES AND WASPS DURING VINTAGE.

OPPOSITE: **SPRINGTIME IN THE VINEYARD AT BLUE MOUNTAIN.**

morning. In 1999, the Krugers bought another five acres (two hectares) near Tuc-El-Nuit Lake at Oliver which had some raw land (since planted) as well as Chardonnay vines. "One of the appealing aspects of that property is that there were no rocks," says Roland, who looks after marketing at Wild Goose.

An engineering designer and also a builder of yachts, Adolf, who was born near Berlin in 1931 and came to Canada as a postwar refugee, planted the vineyard after being laid off by a Vancouver consulting firm. The winery was established in 1990, six years after the vineyard, even though the outlook for British Columbia wines was uncertain. Kruger calculated that he was more likely to succeed making his own wine than selling his grapes. A careful businessman, he also sold grapes to Mission Hill until 1995 while getting his winery's sales estab-

POLYETHYLENE TANKS — AN ECONOMICAL ALTERNATIVE TO STAINLESS STEEL.

miner," Kruger said in 2002. "It could have been one of ours."

In 1992, Larry Gerelus and Linda Pruegger, the owners of Stag's Hollow Winery, purchased a producing vineyard immediately adjacent to Wild Goose even though many advised them not to locate in the Okanagan Falls area. Gerelus is glad they ignored that counsel. "This site gives us a very wide window for ripening grapes," he has found. It is a seven-acre (2.8-hectare) sun-drenched bowl with sandy soil on one section and gravel on the other. None of it is overly fertile. It matures grapes suited to the winery's bold style: plum-rich Merlot and strapping Chardonnay. "We are looking at making a Chardonnay that is big on everything," says winemaker Michael Bartier, who joined Stag's Hollow in 2002.

Before becoming a wine grower, Gerelus, who was born in Winnipeg in 1962, had been an actuary. His partner, Linda Pruegger, had been a financial consultant. For several years after buying the vineyard, they commuted from their jobs in Calgary before opening the winery in 1996. The vineyard was growing Chasselas and Vidal when they bought it. The Chasselas vines were grafted to Merlot and Pinot Noir and all but an acre of Vidal was grafted to Chardonnay. Where others use Vidal exclusively in icewine, Gerelus makes it in both a late harvest and a dry style. The 2001 wine was released as 'Tragically Vidal' because birds consumed most of the succulent grapes from the vines. With a wildlife sanctuary near McIntyre Bluff, the Okanagan Falls growers always have to contend with birds. More recently, the native deer population has risen so dramatically that nearly every vineyard now is protected by high deer-proof fences.

Bartier, formerly the assistant winemaker at Hawthorne Mountain, also used the Stag's Hollow cellar to make the debut 2001 Chardonnay for Eaglebluff Vineyard. Only 70 cases of wine was produced — just enough for Rob Mingay, the Vancouver entrepreneur who established Eaglebluff, and his friends. In 2002, using the cellars of Township 7, Bartier made more Chardonnay, along with Merlot and Pinot Gris, all grown on the 10-acre (four

lished. Since its 1989 debut vintage of 500 cases, Wild Goose has increased its production tenfold.

When he was planning the vineyard, Kruger was advised that neither Riesling nor Gewürztraminer were likely to grow successfully. He persisted, even after a sharp frost in the 1985 winter killed a substantial number of young vines. If he were planting today, he says, he would take a chance on Merlot and Pinot Noir — varieties that Wild Goose has made from purchased grapes. However, he is still satisfied with the choices he made two decades ago. "When you look at Riesling, it is a truly amazing grape," Adolf says. "It is so versatile." The winery makes both an off-dry and a dry Riesling, even uniquely barrel-aging a dry Riesling in 1996. The Gewürztraminer from the Wild Goose site is powerful and spicy with 13 percent alcohol. "We were in Alsace last year and we stayed at a winery where they had Gewürztra-

HAGEN, ADOLF AND ROLAND KRUGER
WILD GOOSE VINEYARDS

LARRY GERELUS AND LINDA PRUEGGER
STAG'S HOLLOW WINERY

FRANK SUPERNAK
BLASTED CHURCH VINEYARDS

MATT AND IAN MAVETY
BLUE MOUNTAIN VINEYARD & CELLARS

hectare) Eaglebluff Vineyard. Mingay expects to have his winery license in time to release wines in the fall of 2003. Because Mingay is husbanding his resources, the winery at this time has no tasting room and does not offer tours. "You need 40 acres just to pay for the tractor," Mingay points out.

Born in Toronto in 1952, Mingay has lived in British Columbia since 1974 except for a short stint in Ottawa as an assistant to Ed Broadbent, then leading the federal New Democrats. In Vancouver Mingay's businesses ranged from founding the Whistler Brewing Company to running a political lobbying firm. He became serious about wine in the 1990s after joining a dedicated group of amateur winemakers. He became so passionate about wine and food that he took a six-month course for chefs. "I enjoyed it immensely," he says. "Then I was looking for a way to put my skills to use and I concluded that a vineyard would be a good place to start." In 2000 he purchased the property on Rolling Hills Road, called Vaseux View Vineyards by the previous owner because of the panoramic view southwest over the lake. Consulting viticulturist Valerie Tait helped Mingay reshape the vineyard to his objectives. Eaglebluff will focus very tightly on three varieties — Chardonnay, which was already on the property, and Pinot Gris and Merlot, which have been planted. There is also mature Riesling in the vineyard; the grapes are being sold. "I don't have a lot of passion for the variety," Mingay admits.

At Hawthorne Mountain Vineyards, there is a lot of passion for Riesling but even more for Gewürztraminer; plantings of this variety now command two-thirds of the winery's 100-acre (40-hectare) vineyard. At times, this is still called the SYL Vineyard because of a sad and perhaps apocryphal tale. In

BRINGING IN THE FRESHLY HARVESTED GRAPES AT STAG'S HOLLOW.

OPPOSITE: **AT THE ZELLER VINEYARD NEAR NARAMATA, ORANIENSTEINER GRAPES AWAIT THE ICEWINE HARVEST.**

1919, so the story goes, Major Hugh Fraser, a Montreal native, brought a delicate English war bride to live on this remote farm that was homesteaded at the edge of the Okanagan wilderness in 1902 by Sam and George Hawthorne. The brothers started building the house with the stone façade that is the winery tasting centre. The major completed it, creating a comfortable home and adding touches of luxury in later years. (It is said to have been the first home in the Okanagan with coloured toilet paper!) The major's bride did not stay to see any of this, departing months after she arrived. She left a note signed "S.Y.L" for "see you later." The dog-loving major farmed the property for decades in the company of his collies, a dozen of which are remembered on individual headstones near the winery.

A vineyard was started in 1961 when Foch and Chelois grapes were planted for Andrés Wines. A businessman named Albert LeComte bought the farm in 1983 and opened the LeComte Estate Winery three years later. The vineyard then was only 21 acres (8.5 hectares), some of it planted to an obscure labrusca variety called Buffalo until LeComte's inspired replacement of it with Gewürztraminer. This is a good site for cool-climate grapes because its northeastern slope rises to an elevation of 1,759 feet (536 metres), making it the highest vineyard in the Okanagan. After LeComte sold the winery in 1996, it was renamed for nearby Hawthorne Mountain. The new owners — Harry McWatters and Vincor International — vastly extended the vineyard, planting Chardonnay, Ehrenfelser (for icewine), Pinot Blanc, Pinot Gris and Pinot Noir. Future plans call for a new winery and expanded hospitality facilities that make the most of the breathtaking view from the vineyard. The view comes with a small price: the three-mile (five-kilometre) ascent from Okanagan Falls involves a narrow road with switchbacks as it climbs from the valley. Five minutes by car today, it was an eternity to the major's bride.

The mundanely named Eastside Road skirts the eastern shore of Skaha Lake between Okanagan Falls and Penticton. Another slow, winding road used primarily by lakeside residents, it got onto wine-touring maps in 2000 when the Prpich Hills winery — now renamed Blasted Church — opened three miles (five kilometres) north of Okanagan Falls. Over the previous quarter-century, Dragan (Dan) Prpich, who grew up in rural Croatia, had transformed an orchard into a 48-acre (19.4-hectare) vineyard. The property is on a windy, west-facing plateau, rising at the back toward steep mountains on the east; the vineyards and the village of Kaleden are directly across Skaha Lake. Some of this panorama is visible through the windows of the tasting room, a modern log cabin. It sits above the two-level 10,000-square-foot (929-square-metre) winery cellar, which Prpich, taking advantage of a high point on the property, dug into the earth. The winery and vineyard have the capacity for

BLOOD-RED JUICE IS PUMPED OVER THE SKINS OF FERMENTING MERLOT AT HAWTHORNE MOUNTAIN TO EXTRACT EVEN MORE COLOUR.

10,000 cases of wine a year but only modest volumes were made prior to 2002, when Prpich retired and sold the property to Chris and Evelyn Campbell.

The Campbells are new to the wine business but are ambitious to have an impact. Chris Campbell, born in Vancouver in 1955, holds a diploma in hotel management from the British Columbia Institute of Technology. German-speaking Evelyn, born in Montreal in 1956, studied hotel management in Salzburg, Austria. They met while working at Vancouver's Bayshore Hotel. Subsequently, both moved into financial careers. She became a certified general accountant, first with an independent practice and later as controller of a software company. Chris, who also studied accounting, became a stock broker in 1982. After deciding they wanted their own business, they spent two years shopping for a winery (and looked closely at five) before Prpich came along.

The Prpich wines had almost no profile because the winery was so new and still small. The Campbells relaunched the winery as Blasted Church Vineyards. The inspiration was an old church in nearby Okanagan Falls that had been moved there in 1929 from Fairview, the abandoned mining town near Oliver. The movers found it very difficult to dismantle the church because the nails were so firmly held by the timbers. Consequently, they set off a controlled dynamite blast inside the church — loosening the nails but damaging the steeple.

At least eight different wines were released under the label because Dan Prpich planted at least 14 different vinifera varieties, ranging from Cabernet Sauvignon to Perle of Csaba. The mainstream varieties from Blasted Church are expected to be Merlot, Pinot Noir, Pinot Gris, Chardonnay and Cabernet Sauvignon. The Chardonnay is primarily the exotic Musqué clone, remarkable for its Muscat aromas and flavours. To restructure the vineyard, the Campbells recruited veteran winemaker Frank Supernak, who had just completed a similar task at Hester Creek. Tragically, Supernak, who was just 41, died in November 2002, just after completing the Blasted Church vintage, in an accident while consulting at Silver Sage Winery near Oliver. Victor Manola, the owner of Silver Sage, had been overcome by carbon dioxide while leaning into a tank of fermenting wine. Supernak jumped in to rescue Manola and was also overcome.

THE WINERIES

Blasted Church Vineyards
378 Parsons Road, Okanagan Falls, BC, V0H 1R0
Telephone: 250 497-1125
www.blastedchurch.com

Blue Mountain Vineyard & Cellars
RR1, S3, C4, Okanagan Falls, BC, V0H 1R0
Telephone: 250 497-8244
www.bluemountainwinery.com

Hawthorne Mountain Vineyards
Green Lake Road, Okanagan Falls, BC, V0H 1R0
Telephone: 250 497-8267
www.hmvineyard.com

Stag's Hollow Winery
2237 Sun Valley Way, Okanagan Falls, BC, V0H 1R0
Telephone: 250 497-6162
www.stagshollowwinery.com

Wild Goose Vineyards
2145 Sun Valley Way, Okanagan Falls, BC, V0H 1R0
Telephone: 250 497-8919
www.wildgoosewinery.com

THE GOLDEN MILE

"WE'RE SO FAR OUT ON THIN ICE THAT, IF YOU TAKE THE AVERAGE
WINTER TEMPERATURES FROM WAY BACK WHEN, EASILY HALF THE
ACREAGE [OF GRAPES] WOULD BE TOAST, OR HEAVILY DAMAGED."
—WALTER GEHRINGER
GEHRINGER BROTHERS

Not much remains of the sporadic gold rush that, in the final decade of the nineteenth century, operated in this part of the Okanagan, now known as the Golden Mile. Today, only a memorial recalls Fairview, the town spawned by gold. It was abandoned once the short-lived gold veins were exhausted. The mines often produced spectacular grades and the Tin Horn Creek Quartz Mining Company briefly traded on the Toronto Stock Exchange. The remains of Tin Horn's mill are just beyond the Gewürztraminer vines at Tinhorn Creek Vineyards, a hilltop winery that opened in 1994, almost a century after the gold mine closed. Tinhorn and the other wineries along the extended length of the Golden "Mile" have been operating longer than most of the gold mines.

The wine growers insist that this bench on the west side of the valley, with its well-drained clay and glacial gravel, is one of the Okanagan's top vineyard areas. Olivier Combret, whose family came here in 1992 after many generations as winemakers in France, is categorical: "The site cannot be questioned in terms of its capability of producing wines. It is always advantageous to be exposed to the sunrise side rather than to the sunset. In many varieties, this produces more flavours." Walter Gehringer of Gehringer Brothers believes that the soils on the west side of the valley support good vine growth and produce "fruit-driven" wines.

OPPOSITE: **RESEMBLING A SMALL CASTLE, THE GOLDEN MILE WINERY OFFERS SPLENDID VIEWS OF THE SOUTHERN OKANAGAN VALLEY.**

The vineyards on the bench begin on a knoll overlooking Oliver, the self-proclaimed wine capital of Canada because of all the wineries in the vicinity. Oliver began after World War I as the construction camp when irrigation was developed in the arid south Okanagan. Today, a hiking trail that begins near the Fairview memorial and heads south provides a panorama of the vineyards of both the Golden Mile and Black Sage Road.

The first winery at the north end of the Golden Mile, not far from the Fairview memorial, is Bill Eggert's Fairview Cellars. It opened in 2000 and makes only red wines from its six-acre (2.4-hectare) vineyard on a plateau above the second tee of the Fairview Golf and Country Club. (Some of the members regret that they did not buy the plateau before Eggert began planting it in 1993, since a spectacular tee could have been perched up there.) "We're in the hottest area in Canada where grapes can be grown," Eggert maintains. "I have some of the best land for supporting reds. I did not want to waste my land on whites."

Eggert was born in Ottawa in 1957 and raised in northern Ontario, where his father was a mining engineer. An uncle had a vineyard near Beamsville, south of Hamilton. A few summers there fired Eggert's interest in grapes and a taste for wine, always available at his uncle's table. Eggert got an agriculture degree at Guelph and worked in the vineyard until he failed to convince his uncle to grow vinifera. Eggert moved out in 1983 and, over a decade of working in vineyards and at construction, saved the money to buy his Fairview property. He had planned only to grow grapes but opened his

THESE SOUTH OKANAGAN VINEYARDS ARE IN THE GOLDEN MILE, AN AREA NAMED FOR THE GOLD AND SILVER MINES THAT ONCE OPERATED HERE.

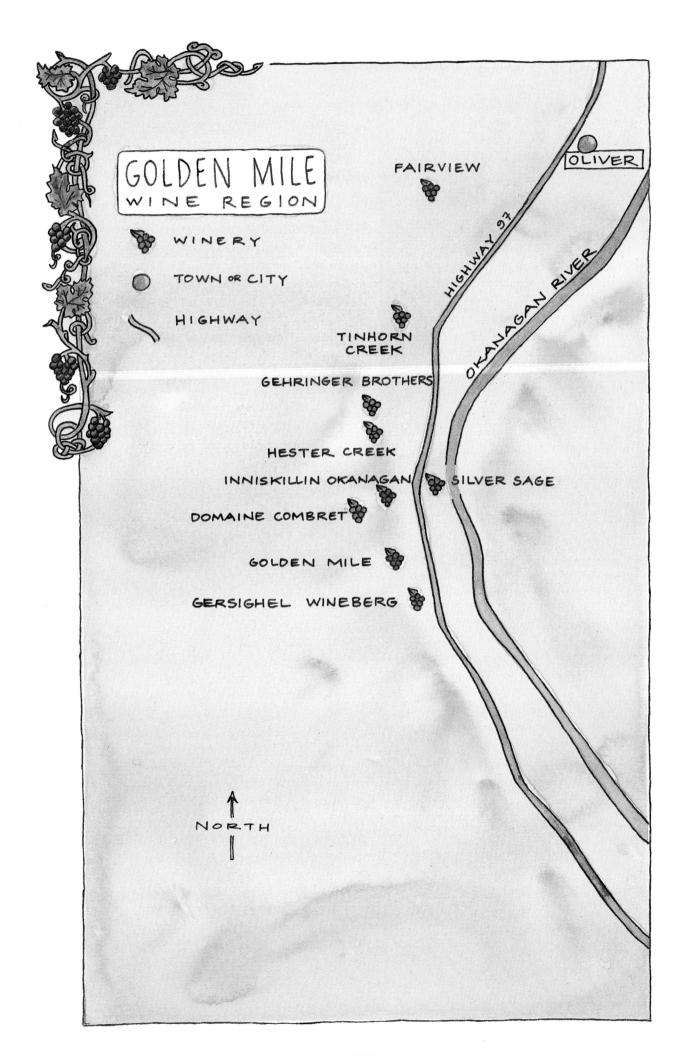

VINES ARE SUPPORTED IN SPRING ON TRELLIS WIRES.

small winery, with a capacity of about 1,500 cases, because wine produces more revenue than grapes.

He was fortunate to launch the winery's hearty reds (Cabernet Franc, Merlot and Cabernet Sauvignon) from the excellent 1998 vintage. The cooler conditions of 1999 yielded wines he considered somewhat lesser in quality. Taking a decision that shows his values, Eggert lowered prices in that vintage; he raised them in 2000 and dropped some in 2001 again. "A lot of it is being able to look people in the face," he says. "I hope the other people in the valley don't keep the prices up in years that are not stellar."

Perhaps it is just coincidental, but the Golden Mile producers seldom price their wines aggressively. Tinhorn Creek, housed in a gold-toned small chateau about 325 feet (100 metres) above the valley, looks expensive but has sold everything except icewine for less than $20 a bottle since opening in 1995. This winery enjoys economies of scale: it has grown to 32,000 cases a year, using grapes from vineyards on both sides of the valley. Its 35 acres (14 hectares) of Golden Mile vineyard, extending even further uphill from the winery was a producing vineyard — mostly Merlot and Chardonnay — when Tinhorn purchased it in 1993. If he were doing it from scratch, general manager Kenn Oldfield says now that he would grow all the Merlot in the winery's 129-acre (52-hectare) Black Sage vineyard, reserving the Golden Mile vineyard for whites. Black Sage vineyards get sun well into the evening, whereas grapes grown on the Golden Mile receive evening shade, a factor contributing to their more vibrant acidity. Having vineyards on both sides of the valley provides enviable options to Sandra Oldfield, Tinhorn Creek's California-trained winemaker.

"We've been caught by surprise at how popular our whites are," Kenn Oldfield says. Tinhorn Creek originally planned to produce 60 percent reds, 40 percent whites. Because its red varieties (Merlot, Cabernet Franc and Pinot Noir) tend to yield more wine than whites, the outcome has been more like 70 percent red. The wines are all successful but Oldfield calculates he could sell about a third more white wines — Pinot Gris, Chardonnay and especially Gewürztraminer — if he had them. Gewürztraminer was something of an afterthought when Tinhorn Creek finished planting the Golden Mile vineyard. The winemaking plan focussed on three red varieties and three whites. Not interested in Pinot Blanc, Sandra Oldfield suggested Gewürztraminer as the third white. The variety matures slowly in the evening shade of the winery's upper hillside vineyard, yielding wines that are crisp and lively.

"At five o'clock on a summer afternoon, we are in the shade of Mt. Kobau," agrees neighbour Walter Gehringer, referring to a nearby peak. "We pick up four hours a day more of active photosynthesis than the east side of the valley would pick up during hot spells." White wines are the specialty at the 18,000-case Gehringer Brothers winery, whose 45 acres (18 hectares) of vines are on the plateau immediately south of Tinhorn Creek. Few wine growers are more methodical than Walter and Gordon Gehringer. The vineyard was purchased in 1981, five years before the winery opened, after the family had done a seven-year climate study to confirm the suitability of the site, which was then growing French hybrid grape varieties. The brothers — born in Canada but trained in German wine schools — planted Riesling, Ehrenfelser and Auxerrois. They only added Bordeaux red varieties, including Cabernet Sauvignon, in the 1990s when they acquired adjoining acreage, christened Dry Rock Vineyard. The Okanagan's last severe winter was in 1983, Walter Gehringer recalls, suggesting that the climate has become warmer. That is why he thinks it is safe to plant tender varieties. "Nobody has to bury a vineyard in its first year anymore," he says. "We did. We covered [the vines] in sawdust and if we had not done that, we would have lost a lot of first-year growth."

Dry Rock has profoundly broadened the winery's initial focus when only Germanic wines were made. The Gehringers planted no mainstream French varieties at first. However, Chardonnay, along with Sauvignon Blanc, Merlot, Cabernet Franc and Cabernet Sauvignon, all mainstays of French winemaking, went into Dry Rock. "[It is] a definite change in direction from where we initially were, allowing us to round out our winemaking style, rather than just making Germanic whites," Walter says. "We've evolved to make wines for certain wine styles, so that I can pretty well boast a wine for anybody's palate. From a fairly narrow portfolio, we have gone to a wider range. I am quite surprised that we won a gold at the [2002 spring] Okanagan Wine Festival for a Chardonnay." He can offer almost two dozen different wines in the

TINHORN CREEK'S VINEYARDS LOOK ACROSS THE MAJESTIC SOUTH OKANAGAN.

BINS OF GRAPES COME IN FROM THE HARVEST.

OPPOSITE: **TINHORN CREEK'S HILLTOP WINERY COMMANDS A DRAMATIC VISTA.**

winery's spacious tasting room because, he believes, climate warming allows him to grow more varieties of grapes. "It's not that we finally know what we're doing," he says humbly. "Somebody else is helping us along the way dramatically and we're then guiding it the rest of the way."

On the next plateau to the south, Hester Creek has some of the Okanagan's oldest vinifera plants in its 71-acre (29-hectare) vineyard. "One of the reasons I was so interested in coming to Hester Creek was the old vines," said Frank Supernak, the winemaker there from 1996 until the summer of 2002, and then at Blasted Church Vineyards until his accidental death in November 2002. "Oldest vines make the finest wines. Eighty-five percent of winemaking starts in the vineyard." Hester Creek was formerly known as Divino Estate Winery until it was acquired in 1996 by a group that included Supernak. The original winery was launched by Joe Busnardo who planted the property between 1968 and 1972 entirely with European grape varieties. "Joe was a pioneer for sure," Supernak said. The big wineries at the time advised against planting tender vinifera. Sure enough, Busnardo's initial 1968 planting of Merlot did not survive the winter. Famously stubborn, he continued planting vinifera, including what Supernak called a "hodgepodge of reds." Divino also planted varieties common in his native Italy, including Garganega, Malvasia and Trebbiano. Of these, Hester Creek retained only Trebbiano, each year making a dessert wine from the variety and, in 2001, a dry table wine.

The vineyard had not been farmed as well as it had been planted. "My biggest challenge was trying to retain most of this gnarly, bush-like vineyard," Supernak said. "It was a tangled web of shoots, cordons and weeds." It took him five years to tame the vines, introduce uniform trellising, replace some rows and fill in between the extraordinarily wide vineyard rows Busnardo had favoured. "But we retained 75 percent of the vineyard," Supernak said with satisfaction.

When Supernak moved to Blasted Church, Hester Creek recruited Kirby Froese as its new winemaker. Born in Moose Jaw, Saskatchewan, in 1970, Froese began as a wine steward at the renowned Fairmont Banff Springs Hotel. In 1993 he spent the harvest at an Australian winery and then returned to North America, learning his trade hands-on with jobs in California, Chile, and in the Okanagan at Sumac Ridge, Hawthorne Mountain Vineyards and Red Rooster. "I've got a New World

BINS OF FRESHLY PICKED GRAPES READY FOR THE CRUSHER.

OPPOSITE: **VIGOROUS LEAF GROWTH IS PRUNED SO THAT THE VINES FOCUS ENERGY ON PRODUCING GRAPES.**

PREVIOUS PAGES: **VINEYARDS AT GEHRINGER BROTHERS ARE ON A PLATEAU HIGH ABOVE THE OKANAGAN VALLEY.**

the stable storage temperatures created by the earth's natural insulation. The winery's Dark Horse Vineyard produces grapes for between 7,000 and 8,000 cases a year, about a third of Inniskillin Okanagan's total annual production. "This vineyard is still very significant," Mayer says. "This is the cream of all the wines that we make."

The Domaine Combret winery, perched high above Dark Horse Vineyard, is a brilliantly efficient design, the first in the Okanagan to harness gravity for moving wines. Wherever possible, juice and wine flows gently by gravity, with minimal use of mechanical pumps. Winemakers agree that gentle handling of the liquids results in superior wines. Many wineries built subsequently in the Okanagan are laid out to reduce pumping to a minimum. Precocious Olivier Combret was newly graduated from the Montpellier wine school in France when he designed the winery for his family, which had been making wine in France for 10 generations. Robert Combret, the young winemaker's father, had spent a summer in the Okanagan in 1958 while completing a master's degree in agriculture at the University of British Columbia. When relocating after selling the family's winery in France, he recalled Okanagan's excellent wine-growing climate.

In 1992, when "everything" was for sale, the Combret family canvassed the valley and bought this Golden Mile vineyard. It was already growing Riesling, Chardonnay and Cabernet Franc and had raw land. It would be, Robert said, "a mere technicality" to make good wines, given the site and the efficiency of the winery. Starting with a silver medal for its 1993 Riesling — easily the most complex Riesling made in British Columbia to that time — the winery has medalled consistently since in international competitions.

While he exudes youthful self-confidence, Olivier argues that the winemaker's influence on wine quality, on a scale of 100, is only 10. The winery design counts for another 10 while the site and the vineyard management contribute 80 percent of the quality. He assigns more credit to the winemaking in difficult vintages. "But we have fantastic conditions in eight of 10 years," he says. The Combret vineyard, with 40 acres (16 hectares) under vine and another 25 acres (10 hectares) available for expansion, has a sun-capturing 10 percent slope to the south. The soil helps impart distinctive mineral notes to the white wines and spiciness to the reds. "Time will tell," Olivier says. "Everyone can make good wines when the vines are young.

GERD DE GUSSEM
GERSIGHEL WINEBERG

KENN AND SANDRA OLDFIELD
TINHORN CREEK VINEYARDS

We still have to wait 25 or 30 years to see the potential of this site."

Before he was a wine grower, Peter Serwo was a builder, as is evident from Golden Mile Cellars, the winery he and his wife, Helga, opened in 1998. Serwo designed and built a small, copper-roofed European castle for the winery, setting it on a height of land overlooking his 25 acres (10 hectares) of vineyards. The winery even has a guardian called Paulus glowering over the parapets. In reality, Paulus is a simulated suit of armour one of the Serwo daughters found at a shop in the British Columbia interior. The baronial ambiance continues beyond the huge doors into the winery's cool tasting room with its lead-paned windows offering views of the vineyard and the valley beyond.

Serwo has been growing grapes on the Golden Mile since 1982. Earlier, he had grown hybrid varieties elsewhere in the Okanagan, including a site in Kaleden too cool to ripen grapes well. On the Golden Mile, a much better growing area, he planted only vinifera, having been encouraged to do so by the late Dr. Helmut Becker, the grape breeder from Geisenheim in Germany. Thus, the wines at Golden Mile are made from such white varieties as Chardonnay, Optima, Ehrenfelser and Riesling and reds like Pinot Noir and Merlot. Most of Serwo's grapes are sold to other wineries; Golden Mile Cellars produces about 1,000 cases of wine annually. Flexible and well-equipped, the winery that resembles a castle can be expanded readily into the hillside. Peter Serwo, 70 in 2002, is likely to leave that expansion for someone else to do.

The Golden Mile bench peters out at or just before it gets to Gersighel Wineberg and its vineyard with Pinot Blanc, Chardonnay and Pinot Noir. This is a small winery beside Highway 97 whose crudely handwritten signs telegraph the breezy informality of owner Dirk De Gussem. A native of Belgium, he opened the winery in 1995, crafting the Gersighel name from the first letters of the given names of each of his three grown children, including Gerd, the winemaker.

Silver Sage Winery, with a 22-acre (nine-hectare) vineyard right beside the Okanagan River on the floor of the valley, shares the soils, if not the aspect, of the Golden Mile. The winery is splendid in design and construction, reflecting the skills of Victor Manola, who established the winery with his wife, Anna, prior to his death in a winery accident in November 2002. He had spent almost two decades building fine homes in Edmonton and Vancouver until, in 1996, he and Anna bought this

OPPOSITE: **PINK-HUED RIPE PINOT GRIS YIELDS SOME OF THE OKANAGAN'S BEST WHITE WINES.**

FRESHLY PICKED PINOT GRIS IS DUMPED INTO THE CRUSHER AT HESTER CREEK.

property. A former producing vineyard that had been fallow since 1988, it has been replanted with Merlot, Pinot Noir, Pinot Blanc, Gewürztraminer and Schönburger.

Both Victor and Anna were natives of Romania. His family owned extensive vineyards and made an indifferent living delivering the grapes and wines they grew to an agency of the Communist government. In 1975, when he was 20 and about to be drafted into the army, Victor Manola had had enough. He escaped to Austria in a circuitous rail journey during which he went without food for 12 days. He came to Canada after a brief stay in a refugee camp. When the Communist government fell, Manola brought not only Anna but his own parents and siblings to Canada. He fell in love with the Okanagan during vacation travels. The Silver Sage Winery, initially making fruit wines for export only, opened in 1999. Two years later, the winery began selling both fruit and grape wines in the domestic market as well. It is clear from some of Silver Sage's blends that Victor Manola was a winemaker who dared to be different. One of his wines is a sweet Pinot Blanc spiced with a chili pepper in each bottle.

THE WINERIES

Domaine Combret
32057 #13 Road, Oliver, BC, V0H 1T0
Telephone: 250 498-6966
www.combretwine.com

Fairview Cellars
Old Golf Course Road (13147 334th Avenue),
Oliver, BC, V0H 1T0
Telephone: 250 498-2211
E-Mail: beggert@img.net

Gehringer Brothers Estate Winery
Road 8, Oliver, BC, V0H 1T0
Telephone: 250 498-3537

Gersighel Wineberg
29690 Highway 97, Oliver, BC, V0H 1T0
Telephone: 250 494-3319

Golden Mile Cellars
13140 316 A Avenue, Oliver, BC, V0H 1T0
Telephone: 250 498-8330

Hester Creek Estate Winery
13163 326th Avenue, Oliver, BC, V0H 1T0
Telephone: 250 498-4435
www.hestercreek.com

Inniskillin Okanagan Vineyards
Road 11 West, Oliver, BC, V0H 1T0
Telephone: 250 498-6663
www.inniskillin.com

Silver Sage Winery
32032 87th Street, Oliver, BC, V0H 1T0
Telephone: 250 498-0310
www.silversagewinery.com

Tinhorn Creek Vineyards
32830 Tinhorn Creek Road, Oliver, BC, V0H 1T0
Telephone: 250 498-3743
www.tinhorn.com

THE BLACK SAGE ROAD NEIGHBOURHOOD

"IF IT WASN'T FOR THE WINE INDUSTRY, THIS SOUTH END
OF THE VALLEY WOULD BE DEADER THAN A DOORNAIL."

—RICHARD CLEAVE
VETERAN VINEYARD MANAGER

No one knows the Black Sage Road vineyards better than Richard Cleave, now completing his third decade as a vineyard manager here. Born in England in 1946, he came to Canada in 1972, armed with an agriculture degree and a few years experience as a farm manager. He worked at a farm at Grand Forks in the British Columbia interior until 1975 when he became manager of Shannon Vineyards, one of the early plantings on this sandy bench south of Oliver.

The viticultural landscape has changed enormously since Cleave arrived. During his first dozen or so years here, almost all the grapes being grown were hybrids and, as was the universal practice in that era, the vineyards were managed for quantity, not quality. Cleave admits that he seldom even consumed the wines made from the hybrids he managed. Nearly all those vines were pulled out after the 1988 harvest; a local feedlot operator grew alfalfa there for the next five years. Then in 1993, the Sumac Ridge Estate Winery planted 115 acres (46 hectares) of vinifera on Black Sage Road. "It was a huge, huge gamble," says Cleave. However, it has succeeded, with the result that Cleave and partner Robert Goltz look after 1,300 acres (526 hectares) of Black Sage vineyard for several wineries. Cleave is now proud of his wine cellar, which includes wines made by Howard Soon under the Sandhill label from

OPPOSITE: **THE WELL-GROOMED BURROWING OWL
VINEYARD IS ON THE STORIED BLACK SAGE ROAD.**

Cleave's intensively farmed eight-acre (3.2-hectare) Phantom Creek Vineyard.

The Black Sage Road vineyards are on what Sumac Ridge's Harry McWatters calls "three hundred feet of beach sand." There are few places in British Columbia where vines grow in leaner soil. In its natural state, the arid earth hosts wiry grasses, tumbleweed and Okanagan rattlesnakes. But with irrigation and fertilizers, the Black Sage Road vineyards produce superb grapes. "We've only found one grape that we can't grow very well here and that's Nebbiolo," says Cleave. "We're growing 21 or 22 different varieties right now and over the next 10 or 12 years, that'll come down to eight or 10 varieties. What we have to do is find out what we can do best. It's a unique grape-growing area in the world. It seems like almost everything we've touched has come through."

At some point in prehistory, this sandy bench on the eastern side of the Okanagan Valley was the shore of a lake. It is believed that glaciers extended to an ice dam at McIntyre Bluff, north of Oliver and south of Vaseux Lake where the valley pinches together. About 60 percent (and rising) of all British Columbia's grapes are grown between McIntyre Bluff and the 49th parallel. South of the bluff on the east side of the valley, the pioneering Inkameep vineyard grows in the rock-strewn soil spilled by the retreating glacier. The soil is so sandy farther south along the Black Sage bench that, as McWatters notes, it squeaks when rubbed between the hands. It becomes only slightly more nutritious as the valley approaches Osoyoos and then crosses into the United States. While the soils are similar, the vineyard attributes of the Osoyoos Lake Bench

AT HISTORIC INKAMEEP VINEYARDS, GRAPES ARE CRUSHED FOR SEVERAL WINERIES.

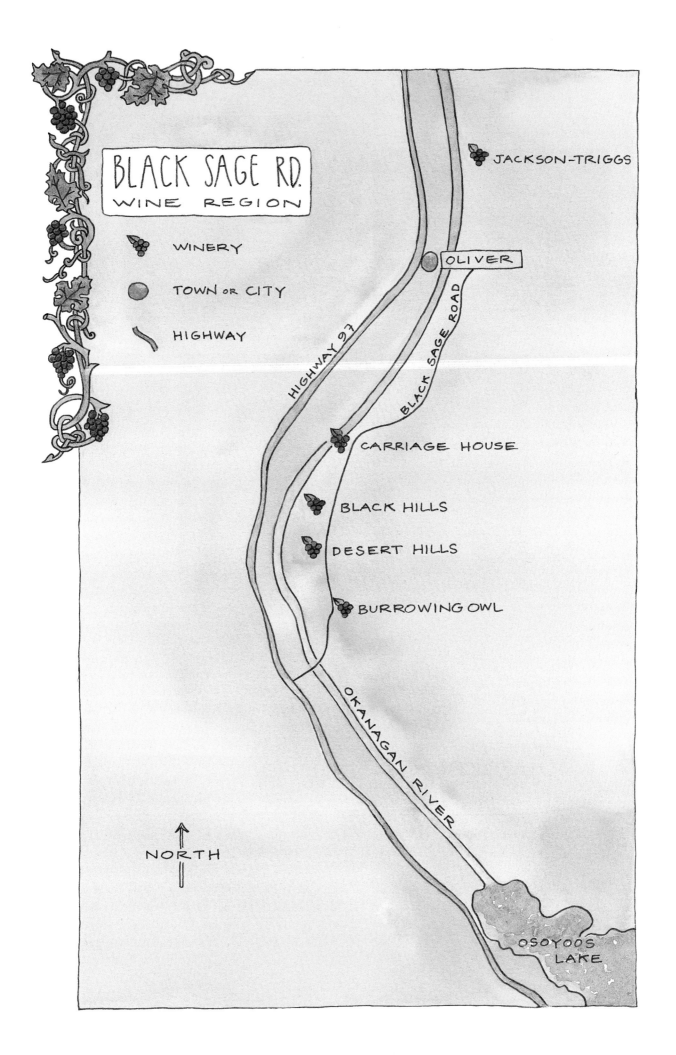

ANNA AND VICTOR MANOLA
SILVER SAGE

JIM WYSE, MICHELLE YOUNG, STEVE WYSE AND MIDGE WYSE
BURROWING OWL ESTATE WINERY

SENKA TENNANT
BLACK HILLS ESTATE WINERY

DAVE WAGNER
CARRIAGE HOUSE WINES

differ significantly from Black Sage Road because of the moderating effect of the lake, absent on Black Sage.

The west side of the valley, by contrast, has heavier soil, with gravel and even a bit of clay mixed in. While the Bordeaux and Rhone grape varieties do well on the eastern side of the valley, the best varieties on the west include Pinot Noir, Riesling and Gewürztraminer. "They grow some damn good grapes over there too if they do their job right," Cleave concedes. The valley, bordered on both sides by mountains, is less than three miles (1.2 kilometres) wide. There are few vineyards on the floor of the valley because it is prone to frost. "The cold air is going to come down from Mt. Kobau as long as there is snow up there," Cleave says, referring to one of the highest peaks on the Okanagan's western border. "It's going to come down every night and push out the warm air. I've been up there in the third week in July — and frozen. It has been absolutely bitterly cold up there." One result of this complex geography is that the opposing slopes of the south Okanagan are different but complementary wine regions.

"It's a really unique little area," Cleave repeats of Black Sage. "We get very little rainfall, especially during the summer months when we get less than three inches on average. We have to use very, very few chemicals. We get few bugs, so we use very few insecticides. It's just a unique area to grow grapes." The sunlight is intense here, perhaps more so than elsewhere because the atmosphere is clean and unpolluted. It is very hot in the summer but prone to frost in spring and in fall when the cloudless sky lets the day's heat escape. "It's radiant frost that's the problem, not cold coming down from the mountain. When you lose that cloud cover, the heat, instead of bouncing back off the clouds, just dissipates into the air."

Severe frosts in winter will penetrate the sand and kill the roots of vines. It has been more than two decades since Black Sage has had a killer winter but Cleave remembers the 1978–79 winter vividly. An early frost blasted vineyards in late September and the vines lost the leaves needed to ripen the fruit. The 1978 crop was huge — 18,404 tons (16,731 tonnes) — an Okanagan record that stood for years. Grapes remained on the vines into November when they froze. Subsequently, vines themselves were damaged; the 1979 harvest was only 10,400 tons (9,455 tonnes). "We learned an awful lot from that freeze," Cleave recounts. "Number one, we don't load individual vines the way we used to. Some growers could grow 40 pounds per vine on the early-ripening hybrids. Now we're down to as little as four to five pounds per vine. That's going to make an immense difference in the quality of what we take off, just by doing that." Most of the vineyards now have extensive frost protection, including powerful wind machines that prevent cold air from gathering in damaging pools.

More than anywhere else in the Okanagan, this looks like wine country, with uninterrupted vineyards tight against each other, laser-straight vine rows rising gently toward the low mountains to the east. The tower at the Burrowing Owl winery is one of the best places from which to view the outstretched vineyards. Until irrigation was brought to the southern Okanagan after World War I, few crops were cultivated here. Once water was available, orchards and gardens were planted and, in the late 1930s, the occasional modest vineyard. But it was only in the mid-1960s, when the big wineries needed grapes, that the very large vineyards were planted in the sandy soils. In 1965 Calona Vineyards teamed up with Richard Stewart (whose sons now run the Quails' Gate winery) to develop what they called Pacific Vineyards. Just to the north of Pacific, the Monashee Vineyard was planted by a school teacher from Kelowna named Ed Wahl and a group of investors. They bought 265 acres (107 hectares) of raw land and turned it into one of British Columbia's single largest vineyards, so large that the Okanagan's first mechanical harvesters were used here. These contiguous vineyards had one thing in common: all were planted with French hybrid grapes because the owners did not believe the European varieties now doing so well would survive the winter.

THE OSOYOOS INDIAN BAND, THE OKANAGAN'S MAJOR VINEYARD LANDLORD, HAS MEMBERS WHO READ A SALISH DIALECT.

In 1971, when the Capozzi family sold Calona Vineyards, the new owners (a Montreal conglomerate) generated cash by selling the vineyard to Harry Shannon and a syndicate of private investors. Both Shannon and Ed Wahl's consortium seriously considered developing wineries on Black Sage Road. Neither had gotten around to it by 1988 when, just prior to free trade with the United States, the provincial government offered vineyards cash to pull out grapes not suitable for premium wines. By the spring of 1989, Black Sage was denuded of almost all vines. The land was sold and very nearly planted with ginseng before, in 1992, confidence returned to the wine industry. Cleave, who remained after the pull-out to run a few residual acres of vinifera, figures: "You could probably have bought this whole bench for $1 million. Now you couldn't buy it for $40 million." That is because, on the heels of the Sumac Ridge planting, the remainder of the old vineyard land was acquired and planted by a variety of owners, including Tinhorn Creek and Burrowing Owl. In 1996 Mission Hill acquired a vineyard planted in 1993 by an aspiring vintner who later changed his mind. All make highly regarded wines.

Andrew Peller, the founder of Andrés Wines, was perhaps the first vintner to understand the potential of Black Sage Road. He wanted to plant a vineyard there when he was starting the winery in 1961. In his autobiography, *The Winemaker,* he wrote that, south of Penticton, he "came across a plateau covering over one thousand acres of rich land absolutely perfect for grapes. A little concentrated research showed me that I could never buy it! The land was set aside as a reservation for the local Indians." The 32,251-acre (13,052-hectare) reservation of the Osoyoos Indian Band is on the east side of the Okanagan River. It extends from Osoyoos Lake to Gallagher Lake and includes a substantial part, although not all, of Black Sage Road's vineyards. The largest one at the northern extension of Black Sage is the Band-operated Inkameep Vineyard just northeast of Oliver, now 240 acres (97 hectares) in size.

This vineyard was begun in 1968 at the urging of Peller. At first hybrid grapes were planted, but in 1975 Andrés imported Riesling, Ehrenfelser and

OPPOSITE: **GRAPES BEING CRUSHED AND DESTEMMED AT THE BLACK HILLS WINERY.**

THE LUSH VINE CANOPY IS PRUNED AT TOOR VINEYARDS ON BLACK SAGE ROAD.

Scheurebe from Germany. These are not the best varieties for a terroir more suited to growing big red wines. At that time, the Okanagan was considered better for growing the German white varieties, whose wines were then in demand by consumers.

The Inkameep Vineyard had a very difficult birth when it began in 1968. A delay in completing the

THE VINEYARD SCARECROW — COLOURFUL BUT INEFFECTIVE IN KEEPING BIRDS FROM THE GRAPES.

irrigation system that summer resulted in 20 percent loss among the water-starved new vines. More vines died in late December when the temperature plunged well below freezing. The vineyard struggled through its first decade, and was even written off as a failure by the Department of Indian Affairs, until it found its feet. The turning point came when Andrés imported the vinifera from Germany. Subsequently, almost the entire vineyard was transformed to one of the single largest plantings of premium grapes in the Okanagan. Today the vineyard grows about $1 million worth of grapes annually and supplies eight or nine wineries, still including Andrés.

In 1981, T.G. Bright & Company of Niagara Falls, a predecessor company to Vincor, opened a winery beside the vineyard. Then Canada's oldest wine company, Brights leased a building which the Osoyoos built. The winery has grown substantially and now is Vincor's Jackson-Triggs winery. It was a contentious project for the Osoyoos people. Sam Baptiste, then the elected chief and now the vineyard manager, drew sharp criticism from other chiefs who were aghast at a winery on reserve land. He nearly called off the agreement. "I thought about it and I decided it was none of their business," he recalled later. Band members in a referendum approved the winery because it offered employment. Clarence Louie, who succeeded Baptiste as chief, now says: "We've had people retire after putting in twenty-plus years of service there. Those winery jobs are good-paying jobs." The Osoyoos today are major players in the wine industry. Vincor has leased and planted nearly 1,000 acres (400 hectares) of reserve land and has an option to lease as much again. The Osoyoos's own winery, NK'Mip Cellars, opened in 2002.

Most of the grapes grown in the Black Sage Road vineyards are processed in wineries located elsewhere in the Okanagan. However, three wineries currently boast a Black Sage address: Carriage House, Burrowing Owl and Black Hills.

Toward the south end of Black Sage Road (also called 71st Street on Oliver maps), is the 23-acre (9.3-hectare) Toor Vineyards, planted in 1996 in part of a former orchard operated by twin brothers Randy and Jessie Toor and their younger brother, David. Born in India, the brothers grew up in Winnipeg, where their mother settled after coming to Canada. In 1988 the family moved to what was then an orchard on Black Sage Road. It soon became Randy's ambition to grow grapes. With help from Domaine de Chaberton winery, which buys premium red grapes from this vineyard to supplement the whites in its Fraser Valley vineyard, the Toor brothers converted part of the property to vines, including Syrah, Cabernet Sauvignon and Merlot. Elias Phiniotis, the consultant who makes wines at Domaine, was enlisted by the Toor brothers. In 2002, with about 1,000 cases of wine ready to market, they applied for a winery license under the name Desert Ridge Estate Winery. By coincidence,

CedarCreek has chosen the same name for its proposed boutique winery at Osoyoos and registered the name first. The Toor brothers switched to Desert Hills and discovered they liked the name better anyway. Whatever his winery is called, Randy Toor is determined to be a small producer of quality wines. "This is my dream," he says.

"We wish there were more wineries on the strip," says Robert Tennant, one of the partners at Black Hills. Black Hills opened in the spring of 2001 with only two wines: 1,600 cases of 1999 Nota Bene, a full and complex red blend based on Merlot, and a mere 140 cases of Sequentia, a late-harvest Sauvignon Blanc. "I like Latin a lot," winemaker Senka Tennant smiles, explaining the proprietary names for the wines. In earlier careers, both Tennant and her husband were teachers in Vancouver. In the mid-1990s, the Tennants and their vineyard partners, Susan and Peter McCarrell, decided to get "back to the land." In 1996, they bought a 34-acre (13.8-hectare) property on the downhill side of Black Sage Road, a site where the sandy soil is mixed with gravel. The vineyard had not participated in the 1988 uprooting and still had hybrid varieties. The new owners promptly replanted with Pinot Noir, Merlot, Chardonnay and Sauvignon Blanc, with the largest block (just over nine acres/ 3.6 hectares) dedicated to Cabernet Sauvignon. A rustic Quonset hut formerly used to build cars for demolition derbies was turned into a winery. "We call it Modern Granville Island architecture," Robert Tennant laughs about the recycled winery building. "We're trying to put our money into our product."

VINEYARDS ON THE BLACK SAGE BENCH FLOURISH IN A DRY DESERT ENVIRONMENT.

BINS OF FRESHLY PICKED CHARDONNAY FROM BURROWING OWL VINEYARDS.

An experienced amateur winemaker, Senka Tennant prepared for her commercial debut with winemaking courses and by retaining consultant Rusty Figgins, a Washington State star winemaker who specializes in Syrah. Black Hills is focussed on a single blended red. "Blends are the way to go," says Tennant, who wants to craft one superior red each vintage rather than release the varietals on their own. The 2000 Nota Bene — which means "take notice" — is a blend where Cabernet Sauvignon takes the lead at 52 percent, supported with Cabernet Franc and Merlot. "We're not trying to make our wine identical from year to year," Robert Tennant explains. "We want people to look forward to tasting the next vintage." Black Hills intends to remain a small, exclusive producer of no more than 2,500 cases a year. There are plans for a dry white table wine made with Sauvignon Blanc in a style influenced by the wines of Sancerre in France. Black Hills also provides grapes to CedarCreek.

Carriage House opened in 1995, making a specialty of Kerner, a white variety much admired by Dave Wagner, the winery's owner and winemaker. He was born in New Westminster in 1952 but grew up in the south Okanagan. He began making fruit wine at home during a brewery strike in 1972. After he moved to grapes, he became particularly enamoured of Kerner, a German-bred variety with Riesling parentage. "I really enjoy the aroma of Kerner fermenting," he says, comparing it to grapefruit and pineapple. "I'm very biased. I can't think of a year when Kerner has let me down." In his tasting room, where he has two styles of Kerner, he proselytizes this variety, which few wineries use, as an alternative to Chardonnay. He points out that consumers wanting Chardonnay have "tons of nice Chardonnays to pick from." He also grows Chardonnay in his eight and a half-acre (3.4-hectare) vineyard but he is considering replacing it with Syrah.

As he has come to know his vineyard, Wagner has concluded that the site favours reds and he seems to waver a bit in his enthusiasm for Kerner. He has, for example, three clones of Pinot Noir. "That's been a very pleasant surprise from this vineyard," he says. "It made me realize that we should be concentrating on reds." He also grows Merlot and Cabernet Sauvignon. "This is Black Sage Road, the best place in Canada to grow reds," Wagner points out. "Why grow whites? The grapes over here have more colour. That is why the wines that come from this side of the valley are deeper in colour." He contends that the Black Sage reds have deeper flavours, reflecting the maturity the grapes achieve during the long, hot days. "I'm letting the vineyard set its own style," Wagner says. "I know what kind of production I'm getting now. I like meaty wines."

The polished wines from Burrowing Owl Estate Winery (named for owls that live in burrows in the South Okanagan) garnered acclaim from the day the winery opened in 1998. This started as a vineyard project only, the dream of Jim Wyse, a successful Vancouver developer who had been seduced by the vineyard lifestyle while working on real estate projects in the Okanagan. "The notion of getting into business only, without a winery, seemed to make sense when I did the numbers," he said. In 1993 he purchased the first 100 acres (40 hectares) of the three contiguous parcels that became Burrowing Owl Vineyards. A few months later, a second parcel, slightly larger, came on the market and Wyse, this time backed by investor friends, bought it as well. The third parcel was added in 1999 when Burrowing Owl leased 72 acres (29 hectares) from the Osoyoos Indian Band. The property then totalled 288 acres (117 hectares), including the winery site.

Wyse was able to sell grapes in his first year as a vineyard owner because the first parcel had producing vines, including 15-year-old Pinot Blanc vines. The rest was planted predominantly with red varieties and, when the vineyards began producing, most of the grapes were sold to Calona Vineyards (largely for its Sandhill wines). By 1996 Wyse observed that Calona was making prize-winning wines with the Burrowing Owl fruit. He figured that Burrowing Owl, by building a modern winery and applying the latest winemaking techniques, could add value by making its own wines as well. Calona's parent company, Cascadia Brands, put up the capital for a 50 percent interest in the winery, while Wyse and his partners contributed vineyards and cash. Shrewdly, Wyse signed consulting winemaker Bill Dyer, who had just left Sterling Vineyards, a leading winery in the Napa Valley, and was looking for the creative challenge offered by Burrowing Owl. His adept winemaking with well-grown grapes contributed to the winery's quick success. By the vintage of 2001, its fifth, the winery was approaching its capacity of

STEVE WYSE CLEANS FERMENTATION TANKS AT BURROWING OWL WINERY.

16,000 cases and was selling it all. In 2002 the ownership was restructured. Wyse bought the other 50 percent of the winery and a third of the vineyard while Cascadia, having scored its own success with the Sandhill wines, took over the rest of the vineyard. "Our whole involvement with Burrowing Owl was to secure grapes for our winemaking program," Cascadia president Bill Baxter says. "In the view of Howard Soon, our winemaker, these are the best grapes in the Okanagan."

While the plantings have included such white varieties as Pinot Gris and Chardonnay, nearly three-quarters of the vines are red varieties, including Syrah, Sangiovese, Pinot Noir and the Bordeaux reds. "Our original big leap of faith was to get into the Bordeaux reds," Wyse says. "The Merlot has definitely been the best red. I always saw this site as one of the hottest and best in the country. It faces the right direction. We are a week or two ahead of the other side of the valley just because of the late afternoon sun." He maintains that wineries on the Black Sage Bench should make what the soil and climate allow them to do best — "and that's the reds."

Dick Cleave has argued for years that red wines are the future for the south Okanagan vineyards. "We were a white industry and we were never going to get anywhere until we had some great reds," Cleave says. "We're getting there. Basically, we are a 10-year-old industry. These vines now in the ground are young vines that aren't giving us their full characteristics yet. There should be a natural progression in quality just from the maturing of the vines."

THE WINERIES

Black Hills Estate Winery
30880 Black Sage Road, Oliver, BC, V0H 1T0
Telephone: 250 498-0666
www.blackhillswinery.com

Burrowing Owl Estate Winery
100 Burrowing Owl Place (off Black Sage Road),
Oliver, BC, V0H 1T0
Telephone: 1 877 498-0620
www.bovwine.com

Carriage House Wines
32764 Black Sage Road, Oliver, BC, V0H 1T0
Telephone: 250 498-8818
E-Mail: carhsewines@otvcablelan.net

Desert Hills Estate Winery
30480 71st Street, Oliver, BC, V0H 1T0
Telephone: 250 498-1040
Telefax: 250 498-3015
www.deserthills.ca

Jackson-Triggs Vintners
38691 Highway 97 North, Oliver, BC, V0H 1T0
Telephone: 250 498-4981
www.atlaswine.com

OSOYOOS LAKE BENCH

"WE STILL HAVE SOME PRIME GRAPE GROWING LAND ON THE OSOYOOS INDIAN RESERVE BUT BECAUSE IT IS ON THE DESERT, WE ARE JUST GOING TO LEAVE IT IN ITS NATURAL STATE. THERE HAS TO BE A BALANCE."
—CHIEF CLARENCE LOUIE
OSOYOOS INDIAN BAND

In the challenging vintage of 1999, Sam Baptiste, the general manager of the Osoyoos Indian Band's Inkameep Vineyards, decided not to allocate grapes for the Nk'Mip Cellars winery, then under development. Baptiste believed the Osoyoos should launch North America's first aboriginal-owned winery only with wines from a top vintage. It was a sound decision. The Okanagan Valley's harvests in 2000 and 2001 were excellent and Nk'Mip opened in the summer of 2002 with 7,000 cases of commendable wine from those vintages.

Nk'Mip is the first winery on the Osoyoos Lake Bench. While most of the winery's grapes are grown at Inkameep Vineyards, 10 acres (four hectares) of Merlot, Cabernet Sauvignon and Syrah were planted around the winery as it was being erected, with similar acreage for subsequent planting. The Osoyoos Lake Bench has only recently emerged as an important Okanagan viticultural region. New vineyards support such producers as Vincor, Mission Hill, CedarCreek and Gray Monk. The area also includes the Osoyoos Larose vineyard, supporting the ambitious joint venture between Vincor and Bordeaux's Château Gruaud-Larose. "Our approach there is to produce an $80, $90 bottle of wine," says Mark Sheridan, Vincor's director of vineyard operations in western Canada. "So we are stopping at no lengths to produce the best quality we can."

OPPOSITE: **MISSION HILL'S PREMIUM VINEYARD ON THE OSOYOOS LAKE BENCH.**

As well as being the 49 percent operating partner of Nk'Mip Cellars, Vincor is the largest grower on the Bench, with two vineyards on the east side of the valley, north of the lake. Its 249-acre (101-hectare) Bullpine Vineyard was planted in 1998 and 1999, predominantly with Merlot, Cabernet Sauvignon and Pinot Noir. The adjacent 163 1/2-acre (66-hectare) Bear Cub Vineyard was planted in 1999, again largely to reds, including 13 acres (5.3 hectares) of Syrah and four acres (1.6 hectares) of Zinfandel. The only whites in these two vineyards are Chardonnay, Sauvignon Blanc and a small but important seven and a half-acre (three-hectare) block of Viognier, a variety from southern France now in fashion in North America. Bullpine is the name of an indigenous conifer. The Bear Cub Vineyard received its name after workers, during the vineyard's development, were confronted aggressively by a bear protecting her cub.

An experienced vineyard professional from Australia, Sheridan joined Vincor in October 1999. His responsibility extends beyond the area, with Vincor having recently planted almost 400 acres (162 hectares) in two vineyards at the Oliver end of Black Sage Road. However, Sheridan rates the Osoyoos Lake Bench as perhaps the best grape-growing land in Canada. "It bathes in the afternoon sun," he says. The vineyards slope southwards toward Osoyoos Lake, whose moderating effect wards off frost in the spring and fall. In the sandy soil, the irrigation can be tailored so that the vines produce small but full-flavoured grapes. "It has a marked effect on red wines," Sheridan says. For the most part, this is red wine country. The Nk'Mip vineyard, the Osoyoos Larose vineyard and Cedar-Creek's nearby Desert Ridge Vineyard grow reds exclusively. The latter two vineyards are on the west side of the valley but also have good aspects. Desert Ridge has a panoramic view over the lake toward the town of Osoyoos.

Unfortunately, Osoyoos Larose currently overlooks a landfill site. It is expected that the landfill will close before Vincor and Groupe Taillon, its French partner, build an elegant winery in the vineyard. In the meanwhile, the production facilities for Osoyoos Larose are in a self-contained mini-winery within Vincor's big Oliver winery. "We've set aside some land on the vineyard," Vincor chief executive Don Triggs says. "But you will see us doing it like Robert Mondavi." This refers to the joint venture to make a premium wine called Opus that was launched in 1979 in the Napa Valley by Mondavi and France's Château Mouton Rothschild. Triggs continues: "We'll make the wine in a separate room entirely [at the Oliver winery] for several years until the vines mature. We'll probably look at building the wine its own home much like Mondavi did with Opus, where I think it was nine years before they actually built the winery. We're looking for the vines to mature before we do that."

While some Osoyoos Lake Bench vineyards are owned privately, most of the new vineyards are on the sprawling Osoyoos reserve. The vineyard potential of the benchlands rising east of the Okanagan River was identified and mapped by Agriculture Canada and the valley's grape growers in 1984 when their *Atlas of Suitable Grape Growing Locations* was published. Since 1998, Vincor, whose lengthy winemaking involvement with the Osoyoos began in 1981, has leased and planted almost 1,000 acres (405 hectares) of the reserve, either on the Bench or on Black Sage Road. Vincor also has the right of first refusal on new vineyard development on a comparable acreage.

While Nk'Mip Cellars is its visible presence on British Columbia's wine scene, the greatest contribution by the Osoyoos has been unlocking large tracts for grape production in one of the few places in the Okanagan where vineyard land was still available. This made it possible for large producers to apply the professional resources to improve the quality of the grapes, both on reserve land and on privately owned vineyards. Mission Hill, which now owns nearly 1,000 acres (405 hectares), has hired university graduates as vineyard technologists and sent some to New Zealand or Australia for training. The winery also provides professional help to its contract growers, so they do not fall behind the improvements happening in company-owned vineyards.

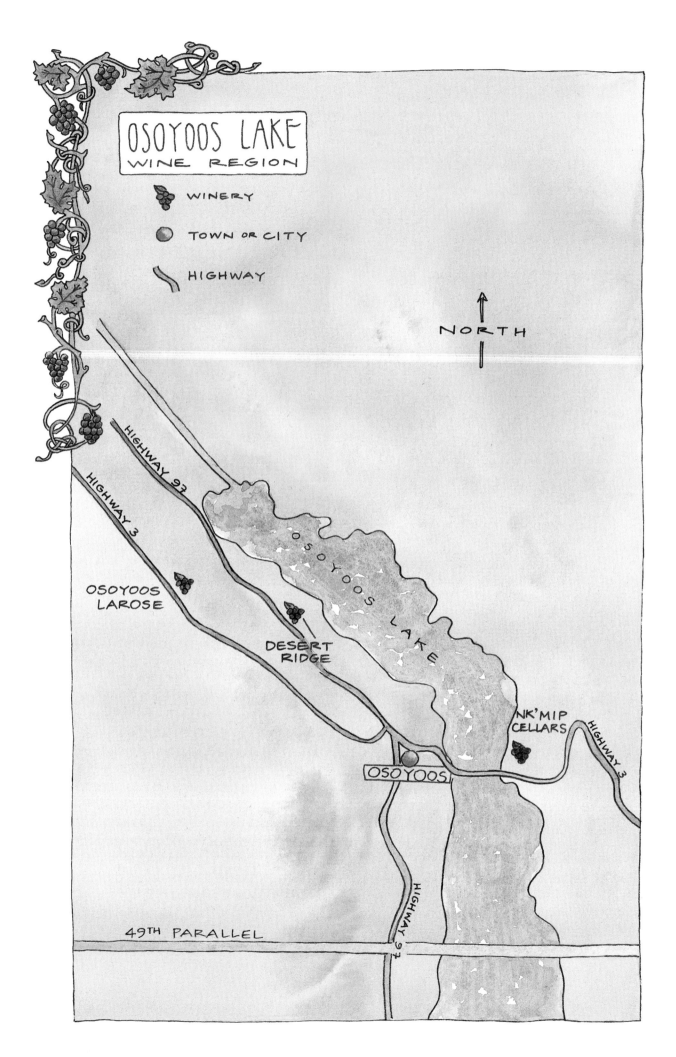

BRUCE NICHOLSON
JACKSON TRIGGS

CHIEF CLARENCE LOUIE
NK'MIP CELLARS

DON TRIGGS
VINCOR

Vincor's Mark Sheridan determined that his first assignment was to advise growers on improving their grapes. "I got here in October 1999," he recounts. "I was here a week and went out to a grower's vineyard and it was overcropped. One side of the vines — it was an east-west planting — one side of the vine was green, one side was ripe. I said 'you'll have to pick this separately.' The guy said 'I can't do that. My family come down every year on this long weekend in October and this is when we pick.'" Over a cup of coffee, Sheridan quietly drew the grower's attention to Vincor's newly planted Osoyoos Lake Bench vineyards, visible from the grower's vineyard, observing that Vincor would be getting a lot of fruit in the future from its own vineyards. "I could see the penny drop. I left and I got a phone call four hours later in which he said he could pick the [green side] separately."

The Osoyoos Indian Band reservation was established in 1877. Only a handful of its 370 members still speak the Salish language, from whence comes the word Nk'Mip. It has been variously anglicized as Inkameep and Inkaneep. The Osoyoos are striving for economic self-sufficiency. Under the Osoyoos Indian Band Development Corporation, they now run ten businesses, including housing developments, a golf course, construction, forestry and sand and gravel companies as well as the winery. Nk'Mip Cellars is the second winery operating on the reserve (the Jackson-Triggs winery at Oliver is owned by Vincor) but the first owned by the band.

Clarence Louie, who was elected chief in 1987 when he was just 27, was away at university when the Oliver winery was built not far from Inkameep Vineyards. "I imagine most vineyard producers have dreams and ideas about opening up their own wineries," Louie says. "The Osoyoos Indian Band is no different, especially since we operate a huge vineyard. This talk about the band getting into its own winery goes back to the 70s." Laid to rest temporarily after the Oliver winery opened in 1981, it arose again toward the end of the decade when the 1989 Free Trade Agreement created deep uncertainty for British Columbia wineries. The general manager of Inkameep then was Kenn Visser. Concerned that grape sales would evaporate if key wineries closed, Visser in 1989 launched Nordique Blanc, a dry white made from Ehrenfelser grapes grown at Inkameep and vinified at an existing Okanagan winery. The market was tested through several vintages but the project was wound up when wineries continued to buy Inkameep's grapes.

It was a failed casino bid in 1997 that led to Nk'Mip Cellars. The Osoyoos Indian Band Development Corporation had created an ambitious destination resort design, including a casino, but the provincial government approved a competing casino proposal. The Osoyoos redesigned the resort with the winery as the anchor. At the urging of Chris Scott, the economic development officer for the band, Vincor was enlisted as the winery's operating partner. Vincor has technical, management and marketing expertise which the 15,000-case winery needs. Equally important, Vincor and the band have become interdependent. "Two-thirds of the employees in the Oliver winery are from the Band," says Triggs, Vincor's chief executive. "Our relationship goes back 25 years. Our winery is on band land. We now have vineyards developed on band land. Our future in the Okanagan is very much intertwined with the future of the band, and when the band expressed an interest in developing a winery of its own, we said we'd be delighted to be their partner." Bruce Nicholson, the award-winning Jackson-Triggs winemaker, made Nk'Mip's wines in 2000 and 2001. Randy Picton, who had been the associate winemaker at CedarCreek, took over just before for the 2002 vintage. Nk'Mip debuted with four varietals, Pinot Noir, Merlot, Chardonnay and Pinot Blanc, and will extend the range as its own vineyard becomes productive.

As North America's first aboriginal-owned winery, Nk'Mip Cellars is a showpiece in every regard. Located on a height of land just at the eastern entrance to Osoyoos, the winery overlooks both the town and the lake. The winery, designed by Penticton architect Robert Mackenzie, is an elegant, Santa Fe–style building. This architecture will be echoed by other proposed resort buildings, including a planned hotel. The project includes a desert interpretation centre and is next to a band-owned campground. "It is a unique setting," Chief Louie says.

THE WINERIES

Nk'Mip Cellars
1400 Rancher Creek Road, Osoyoos, BC, V0H 1V0
Telephone: 250 495-2985
www.nkmipcellars.com

Osoyoos Larose
38691 Highway 97 North, Oliver, BC, V0H 1T0
Telephone: 250 498-4981
www.atlaswine.com

** Desert Ridge Vineyard
14418 Highway 97, Osoyoos, BC, V0H 1V0
Telephone: 250 495-4046

** UNDER DEVELOPMENT AND MAY CHANGE NAME

THE SIMILKAMEEN VALLEY

"IT IS EASY TO GROW A GRAPE BUT TO GROW A GRAPE TO THE HIGHEST QUALITY IN OUR CLIMATE, WITH OUR SOIL TYPES, WITH VINIFERA AND ROOTSTOCK COMBINATIONS, IS MORE CHALLENGING. AND EVERY SITE HAS ITS OWN PERSONALITY."

—VALERIE TAIT
VINEYARD CONSULTANT

Andrew Peller, the founder of Andrés Wines, was the first major winery operator to consider growing grapes in the Similkameen Valley because, to get a winery license, he had to commit to using British Columbia grapes. Peller, who came from Ontario, wrote in his autobiography, *The Winemaker*, that he expected to find abundant vineyards in the Okanagan because of its "perfect climate." He was in for a surprise when he did some homework. "In 1958, only about 250 acres were under grape cultivation, mostly for private use."

Having made a promise, Peller went around the valley, interviewing farmers to determine why so few vineyards existed. He was told that a "viticulturist from Switzerland [had been] notably unsuccessful. Ever since, his failure had been cited as a rule of thumb, with the result that very few others had tried. All my research indicated that this attitude was wrong." He did succeed in getting the then-owner of what is now Hawthorne Mountain Vineyards to plant Chelois and Maréchal Foch, two red hybrid varieties that thrived there for 30 years until they were pulled out in 2001. Given the otherwise substantial disinterest of Okanagan farmers, Peller decided that Andrés would develop its own vineyard. He bought a 40-acre (16-hectare) farm on a bench overlooking the Similkameen Valley; at the time, it was an unsuccessful apple orchard lacking irrigation even though the river was nearby. "One

OPPOSITE: **HOT AND DRY, THE SIMILKAMEEN VALLEY IS STEEPLY BORDERED BY MOUNTAINS.**

of my pieces of research had concerned the benefits accrued by irrigating grapes at specific points in the growth cycle," Peller wrote. "It had been proven that the yield could be increased 30 to 40 per cent with controlled watering."

Peller had high hopes for this vineyard. "The day I went to see the property, the heat nearly

MT. BOUCHERIE'S SIMILKAMEEN VINEYARDS.

melted me," he recounted. "It was the ideal weather for grapes. From this point of view the farm looked like a good investment." Unfortunately, his privately published biography left many loose ends and the fate of this vineyard was one of them. "In the long run, we found it more expedient to rent the vineyard to a farmer and let him operate it," Peller wrote. "He never did make the stipulated payments on time …." The Andrés experience was so unsatisfactory that for decades the winery stayed away from owning its own vineyards. (The property once again has become an orchard.) The winery returned to the Similkameen only in late 1997 to develop its 70-acre (28-hectare) Rocky Ridge Vineyard in partnership with growers Roger and Annette Hol. Rocky Ridge was an alfalfa field when Andrés became involved. Nine different vinifera grape varieties have been planted there, with the biggest plantings being Merlot, Cabernet Franc, Chardonnay, Cabernet Sauvignon and Gamay. The grapes currently are processed in Port Moody, a Vancouver suburb, at the winery which Andrew Peller built in 1961 because Port Moody offered him a site at a very low price. Within the next decade, Andrés is likely to build a new premium winery either in the Okanagan or in the Similkameen.

There has been extensive fruit and vegetable farming here for more than a century. The Similkameen's first commercial orchard was established in 1897 by Francis Xavier Richter, who is remembered today by the Richter Pass, which connects the Similkameen and Okanagan Valleys. While Peller never explained why his apple farm failed as a vineyard, one can surmise from the experience of others. The Similkameen Valley presents its own challenges. The area in which vineyards have been developed, or at least been considered, is the narrow valley extending from Keremeos 11 miles (18 kilometres) to Chopaka on the American border. "Due to the high mountains on both sides of the valley, with lots of rocks, the heat is held in the valley long after the sun sets," says consultant John Bremmer, who once grew grapes here. "This gives the 'oven effect' and is a great place to grow red grapes." The river moderates temperatures in nearby vineyards but not nearly as much as the large lakes moderate the Okanagan's temperatures. Thus, the Similkameen is hotter in summer and colder in winter than the Okanagan. George Hanson, who planted his Harmony-One Vineyard in 2000, recounts one recent summer when, for 43 consecutive days, the noon temperature was above 104°F (40°C). The valley is arid, with persistent winds throughout the year that suck the moisture from the vines and the soil. Hanson, who has overhead sprinklers in his 15-acre (six-hectare) vineyard, calculates that as much as three-quarters of the water evaporates between the sprinkler heads and the ground unless the irrigation is done at night. At Crowsnest Vineyards, strong gusts once ripped the flat roof from the winery and dropped it in a neighbour's yard. The winery was rebuilt with a peaked roof designed to withstand the wind. Both at

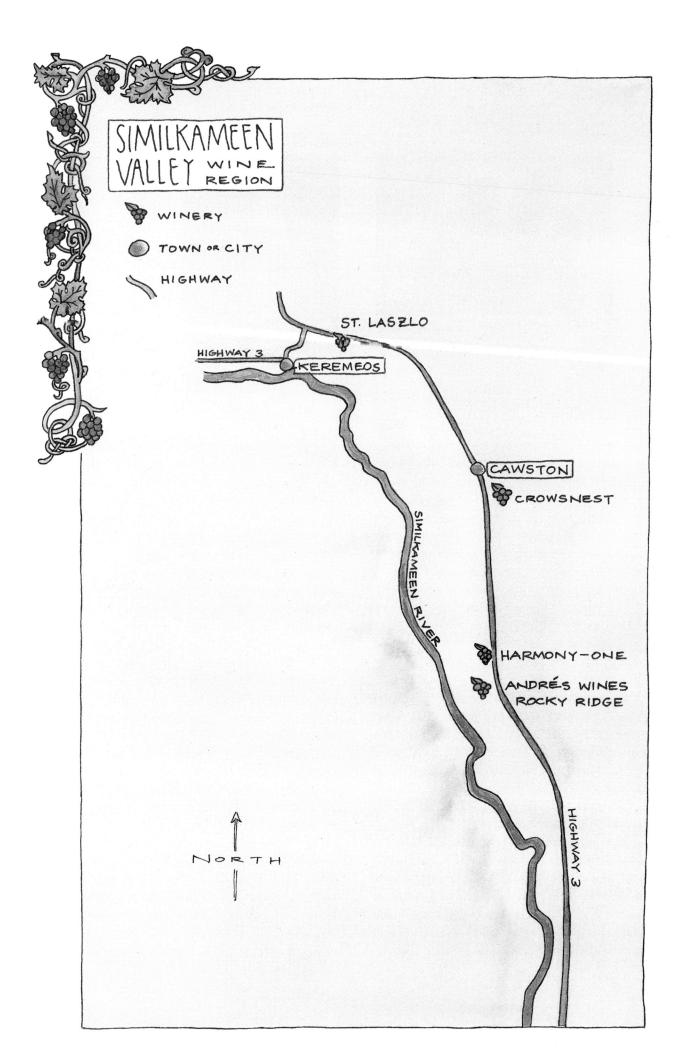

SASCHA AND ANN HEINECKE
CROWSNEST VINEYARDS

JOE RITLOP JR.
ST. LASZLO VINEYARDS ESTATE WINERY

Crowsnest and at Harmony, shelter belts of trees have been planted along the most wind-vulnerable side of the vineyards. Even that has its disadvantages. Sascha Heinecke, one of the owners at Crowsnest, complains that the trees have given predatory birds a convenient roost beside the vines. Hanson argues, however, that the constant wind, in balance, is beneficial. "There are virtually no pests and mildew is not a big problem," he has found. "It is pretty nice when you don't have to worry about pests."

The Similkameen's reputation for killing vines by desiccation and cold arose in a time when growers did not routinely irrigate the vineyards *after* harvest as they do now. Heavily cropped vines (and Peller's reference to increasing yield with irrigation suggests those in his vineyard were) may need extra weeks to ripen the grapes. If the harvest does not occur until very late autumn, the vines have little chance to become dormant before sharp winter temperatures set in. They will suffer fatal damage from the cold if not dormant.

That is precisely what happened in 1978 to a project called Similkameen Vineyards, killing not only the vines but also a promising estate winery project. Robert Holt, who was general manager of Jordan & Ste-Michelle Cellars, partnered with Bremmer, then general manager at Andrés, to acquire an 86-acre (35-hectare) vineyard south of Cawston for their own winery. The vineyard that fall had a prodigious crop, about 400 tons (363 tonnes), and the grapes were not harvested until late in the season. Consequently, the vines did not have a chance to become fully dormant before the temperature plunged well below freezing. "Even with the large harvest of 1978, the vines would have survived if the vineyard had been well watered immediately after harvest," Bremmer says. "The soil was dry and had no moisture from August until the following April. The roots and trunks died from desiccation and not from overcropping or the cold." So many vines were killed that the 1979 crop was only a tenth of the previous year's crop. (Some Okanagan vineyards suffered comparable damage for the same reason: the total grape harvest in 1979 was reduced to about half of the previous year's.) Similkameen Vineyards was economically devastated and the winery plans evaporated. Holt, who purchased Bremmer's interest, ran the vineyard for the next two decades until he sold it to the Gidda Brothers, owners of the Mt. Boucherie winery.

Keremeos Vineyards, now called St. Laszlo Vineyards, was the first winery in the Similkameen Valley when it opened in 1984. The 10-acre (four-hectare) vineyard is on the highway just at the southeastern edge of Keremeos. Slovenian-born Joe Ritlop, who opened St. Laszlo, planted a vast array of grapes at his Keremeos vineyard. St.

Laszlo's rustic tasting room still offers varieties unlikely to be seen on wine labels anywhere else in Canada. Among them are Clinton, an American hybrid, and Interlaken, another hybrid with a Muscat aroma, usually grown as a table grape because it is seedless.

Joe Ritlop Jr., who has taken over winemaking duties at St. Laszlo, has introduced more vinifera at the winery. But he continues to make the wines in his father's style, eschewing winemaking chemicals and relying on natural yeast to ferment the wines, producing wines that are rarely mainstream in style. St. Laszlo also makes fruit and berry wines. "I am the guy that resurrected the fruit wine idea," says Joe Jr. He was responding, he says, to requests from tourists visiting the tasting room after it opened in 1985. "So I accessed some raspberries and made a trial batch," he says. "It proved to be very popular." Now he makes wines from cherries, strawberries, pears, peaches, blackberries, blueberries, saskatoons and even elderberries. Indeed, he is willing to step up to any challenge. In 2001 a rose grower in Keremeos asked for wine from rose petals, delivering him 50 pounds (22.7 kilograms) of petals of assorted hues. The wine is as exotic in aroma and flavour as a Persian bazaar but has a weakly pink hue. "Imagine what I could have done if I had had all *red* roses!" he wondered. In the following autumn, he made a point of using only red petals. "It is much more intense and superior," he says. The fragrant wine is now also on sale in the winery's tasting room.

To date, only three other wineries have arisen in the Similkameen. Crowsnest Vineyards at Cawston, with 13 1/2 acres (5.5 hectares) of grapes, was estab-

RAINBOW TOUCHES DOWN IN THE ST. LASZLO WINERY'S VINEYARD AT KEREMEOS.

lished in 1995 by Hugh and Andrea McDonald. Perhaps because of its comparative isolation from the wine route, it was only making about 500 cases of wine a year when the McDonalds sold it in 1998 to the Heinecke family. The Heineckes expanded aggressively and were producing about 7,500 cases a year by 2002.

Olaf and Sabine Heinecke had emigrated a few years before from Leipzig in the former East Germany. Before German reunification, viticulture there was modest. In any event, the Heineckes did not have the resources to start a winery there even if the East German regime had encouraged the family's entrepreneurship. But soon after the Heineckes arrived in British Columbia in 1995, they acquired a small vineyard near Naramata. They followed that by buying a large one near Penticton and finally by purchasing Crowsnest, where they enlisted their two children to run the winery. Sascha, now 25, had just completed a German degree in hotel management. He agreed to come to Canada in 1999 for a two-week vacation and decided to stay, handling the winery's marketing. His sister Ann, four years his junior, earned a winemaking diploma at Weinsberg in Germany, acquiring the skill to make clean, crisp white wines and solid reds at Crowsnest. When the Heineckes took the winery over, Crowsnest predominantly made white wines, including Auxerrois, Chardonnay, Gewürztraminer and Ortega. Even though Sascha argues that the Similkameen is better for white wines than for reds, he planted red varieties exclusively when he increased the vineyard. The winery thus added full-bodied Merlot and dark, firmly-textured Pinot Noir to its range, with Syrah to be added when those plantings begin to produce.

George Hanson has also pegged the torrid Similkameen as red wine country, planting five reds. He planted Chardonnay reluctantly, calculating that consumers will also expect a white wine. His Harmony-One Vineyards and Winery was still under development in 2003. Hanson's target is 4,500 cases of premium wine a year. The winery's name reflects the poetic streak in Hanson's personality. He spent a long time considering names until it struck him one day that a winery owner is like an orchestra conductor, pulling elements together into harmony.

Born in Alberta in 1957, Hanson spent 25 years in the Yukon, becoming a manager in the territory's telephone system. "I got an early golden handshake from the telephone company and decided to pursue my dream," he recounts. His interest in wine, he says, began when a brother married into an Italian family that included a father who was a good winemaker. Hanson learned how to make wine. He also began travelling to wine regions to further his growing passion. "I thought it would be a nice thing to do when I retired at 55," he says. "It promises a good lifestyle." The unexpectedly early severance from the telephone company, more than a decade before he intended to retire, enabled him to begin looking for vineyard property. After several unsuccessful bids on property near Oliver, he chanced on his Similkameen property, then a hay meadow, on the very day in 1999 when the owner put up the sale sign. Hanson bought it immediately — and just ahead of two wineries that were also interested. It is well located beside the highway, on a westward-sloping plateau with panoramic views of the mountains across the valley. A natural gully on the property will house a winery with gravity flow for handling the wines gently. "I'm positioning everything for high quality," Hanson says. "I want to make something beautiful. I want to make a difference."

Hanson also wants to anchor the Similkameen appellation in the grapes grown there. It is not a new idea. When Andrew Peller was trying to get a vineyard established in the valley, he also launched a brand called Similkameen Superior. The wine, which remains in the Andrés portfolio, seldom has had Similkameen fruit in it. Perhaps the grapes from their Rocky Ridge vineyard will change that.

THE WINERIES

Andrés Wines Rocky Ridge Vineyards
2120 Vintner Street, Port Moody, BC, V3H 1W8
Telephone: 604 937-3411
www.andreswines.com

Crowsnest Vineyards
Surprise Drive, RR1, S18, C18, Cawston, BC, V0X 1C0
Telephone: 250 499-5129
www.crowsnestvineyards.com

** Seven Stones Winery (formerly Harmony-One Vineyards and Winery)
1143 Highway 3, Cawston, BC, V0X 1C0
Telephone: 250 499-2144

St. Laszlo Vineyards Estate Winery
Highway 3, Keremeos, BC, V0X 1N0
Telephone: 250 499-2856

** UNDER DEVELOPMENT

INDEX

A'Very Fine Winery 53
Adora Estate Winery 102
Agria 30, 31, 44, 56
Alderlea Vineyards 18, 30
Andrés Wines 59, 118, 166, 181, 182, 186
Apple cider 26, 83
Armstrong, Gary and JoAnn 68
Arrowleaf Cellars 86
Auxerrois 27, 85, 90
Averill Creek Vineyards 21
Avery, David and Liz 53

Bacchus 50
Balsamic vinegar 29
Baptiste, Sam 168, 175
Bartier, Michael 136
Bauck, Wade 57
Bear Cub Vineyard 176
Becker, Helmut 85, 156
Bella Vista Vineyards 67–68
Benchland Vineyards 112–13
Betts, Michael 18
Bieker, Terry 57
Bissett, Lorraine 58
Black Hills Estate Winery 169–70
Black Sage Road 161–72
Blasted Church Vineyards 138–40
Blossom Winery 58
Blue Grouse Vineyards and Winery 29

Blue Mountain Vineyard & Cellars 132–35
Bonaparte Bend Winery 68–69
Bradner, Patrick 70
Bremmer, John 182, 184
Brennink, Albert and Edward 31
Bryden, Tom 69
Bullpine Vineyard 176
Burrowing Owl Estate Winery 82, 170–72
Busnardo, Joe 17, 148

Cabernet Sauvignon 30, 44, 120, 150, 169
Calliope Vintners 118, 120
Calona Vineyards 81–82, 165
Campbell, Chris and Evelyn 140
Carriage House Wines 170
Casabello 107
Cascadia Brands 82, 171, 172
CedarCreek Estate Winery 80–81, 108, 169, 176
Chalet Estate Vineyard 18
Chardonnay 45, 55, 96, 136, 147
Chase & Warren Estate Winery 33
Chasselas 79, 90
Chateau Wolff 18
Cherry Point Vineyards 26, 29
Cipes, Stephen 79–80
Cleave, Richard 161
Coleman, Corey and Gwen 52

Columbia Gardens Vineyard & Winery 69
Columbia Valley Classics 56
Combret, Olivier 143, 155
Cowichan Valley 17, 26
Crowsnest Vineyards 182–86

D'Angelo, Salvatore 120–21
D'Angelo Vineyards 120
Dark Horse Vineyard 150
Desert Hill Estate Winery 168, 176
Divino Estate Winery 17, 148
Docherty, Janet 26
Domaine Combret 155
Domaine de Chaberton 49–52, 168
Dosman, Roger 17, 30
Dry Rock Vineyard 147
Dulik, Susan 73
Duncan Project 27
Dyer, Bill 171

Eaglebluff Vineyard 136
East Kelowna Cider Company 83
Echo Valley Vineyards 31
Eggert, Bill 144–46
Elephant Island Orchard Wines 122
Elmes, Garron 121–22

Fairview Cellars 144–46

INDEX

Ferguson, Bob 126
Fitzpatrick, Senator Ross 81, 108
The Fort Wine Company 57
Fraser, Alex 106
Fraser, Dennis and Pamala 106
Fraser Valley 49–58
Free trade agreement 12
Froese, Kirby 148
Fruit wines 25, 56, 57, 69, 122, 185

Gardner, Paul 112
Garry Oaks Estate Winery 37, 45
Gebert, Leo and Andy 78
Gehringer Brothers Estate Winery 143, 146
Gehringer, Walter and Gordon 146–48
Gerelus, Larry 136
Gersighel Wineberg 156
Gewürztraminer 66, 105, 136, 138, 143, 146
Gidda Brothers 90, 184
Glen Echo Farm 41
Glenterra Vineyards 29
Glenugie Vineyard 53
Godfrey-Brownell Vineyards 30
Godfrey, David 30
Golden Mile 143–59
Golden Mile Cellars 156
Gray Monk Estate Winery 83–84
Greata Ranch 106, 108
Growers' Wine Company 22, 77

Hainle, Tilman 108–109
Hainle Vineyards 108
Halladay, Del and Miranda 122

Hanson, George 182, 186
Harkley, Bill and Janice 45, 46
Harmony-One Vineyards and Winery 182, 186
Harper, John 27, 50
Hawthorne Mountain Vineyards 137–38
Heinecke family 184, 186
Heinecke, Sascha 184, 186
Heiss, George and Trudy 83
Hester Creek Estate Winery 148
Hillside Estate 121
Himrod 78
Holman, Lynn and Keith 116
Honeymoon Bay Wild Blackberry Winery 26
Hooper, Ashley 97–98
House of Rose 78
Huber, Walter 109
Hughes, J.W. 73
Hunting Hawk Vineyards 67

Inkameep Vineyard 166–68, 178
Inniskillin Okanagan Vineyards 150

Jackson-Triggs Vintners 168, 178
Johnston, Andrew 21
Jordan & Ste-Michelle Cellars 23, 73, 184

Kasting, Arlene and Norman 41
Kelly, John 17, 29
Kettle Valley Railway 105, 111
Kettle Valley Winery 126
Kiltz, Hans 29
Klokocka, Vera 120–21

Kompauer, Paul 53
Kozak, Elaine 37, 45, 46
Kruger, Adolf, Roland and Hagen 135

La Frenz 111, 116
Lake Breeze Vineyards 121–22
Lang, Günther 122–23
Lang Vineyards 122–23
Larch Hills Winery 62–65
Lauzon, Ken 121
LeComte Estate Winery 138
Léon Millot 45
Leroux, Christine 106, 122
Liptrot, Robert 33
Loganberry 25
Long Harbour Vineyards 45
Louie, Clarence 168, 178–79
Lyman Estates 66

Madeleine Angevine 52, 65
Madeleine Sylvaner 52
Mahrer, Beat and Prudence 125–26
Malahat Estate Vineyard 18
Manola, Victor 5, 140, 156
Maple syrup 125
Maréchal Foch 69, 85, 97
Marks, Alan 92
Marley Farm Winery 17
Marley, Michael 17
Martin, Jeff 14, 96-97, 111, 116–17
Mavety, Ian, Jane and Matt 132
Mayer, Sandor 150
Mayne Island 37, 41
McCarrell, Susan and Peter 169

189

INDEX

McIntyre Bluff 131, 162
McWatters, Harry 52, 107, 138
Mead 33, 69
Mercier, Marcel 37, 45, 46
Merlot 80, 136, 146, 148, 172
Merridale Ciderworks 26
Michurnitz 90
Mingay, Rob 136–37
Mirko, Ross and Cherie 106, 112, 118
Mission Hill Family Estate 92, 128, 176
Moffat, Merna 26
Monashee Vineyard 165
Morning Bay Farm 40
Moss, Kelly 82
Mount Boucherie 89
Mt. Boucherie Estate Winery 90, 184

Napa Valley 12, 52, 57, 92, 176
Naramata Bench 111–28
Nevrkla, Hans and Hazel 62
Newton Ridge 18, 32
Nichol, Alex and Kathleen 126–27
Nichol Vineyard & Farm Winery 126–27
Nicholson, Bruce 179
Niles, Russ and Marnie 66–67
Nk'Mip Cellars 175, 178–79
Nordique Blanc 179

Oldfield, Kenn and Sandra 146
Oliver 144
Ortega 18, 27, 32, 65
Osoyoos Indian Band 166, 175

Osoyoos Lake Bench 175–79
Osoyoos Larose 175–76
Overbury Farm 38, 41

Pacific Agri-Food Research Centre 101
Pacific Vineyards 165
Page, Larry 37
Pandosy, Father Charles 74
Paradise Ranch Vineyards 128
Passmore, Larry 67
Peller, Andrew 166, 181–82
Pemberton Valley Vineyard 70
Pender Island 37, 40
Pentâge Winery 112
Phantom Creek Vineyard 162
Phiniotis, Elias 55, 168
Picton, Randy 179
Pinnock, Chris 21
Pinot Blanc 85, 102, 135
Pinot Gris 27, 85, 135
Pinot Noir 18, 38, 52, 53, 97, 107, 126
Pinot Reach Cellars 73
Pinotage 121
Pipes, Rick 26
Poplar Grove Winery 118
Port Alberni 33
Prpich, Dan 138
Pruegger, Linda 136

Quails' Gate Estate Winery 74, 79, 89, 97–98

Recline Ridge Vineyards & Winery 61, 65
Red Rooster Winery 125

Riesling 73, 136, 155
Ritlop, Joe 184
Ritlop, Joe Jr. 185
Rittich, Virgil and Eugene 77–78
Rivard, Dominic 57
Rocky Ridge Vineyard 182
Rose petal wine 185
Rose, Vern 78
Ross, David and Theressa 83
Rotberger 85

Saanich Peninsula 17, 25
Salmon Arm 61, 62
Salt Spring Island 37, 38, 45
Salt Spring Vineyards 45
Sandhill 82–83, 161, 172
Saturna Island 37, 38
Saturna Island Vineyards 38
Scherzinger, Edgar 105
Scherzinger Vineyards 101, 105–106
Schmidt, Lloyd 74, 107
Schulze, Marilyn 29
Serwo, Peter and Helga 156
Sheridan, Mark 175–78
Siegerrebe 65
Silver Sage Winery 5, 140, 156
Simes, John 93
Similkameen Valley 90, 181–86
Skaha Lake 112, 138
Slamka Cellars 90
Slamka, Peter 90
Smith, Bruce 45
Smith, Fraser 18
Smith, Michael and Susan 61, 65, 66
Soon, Howard 82, 161, 172

190

INDEX

Sovereign Opal 83, 102
Sparkes, Colin 38, 44
Sparkling wine 29, 52, 80, 107–108, 125
Spiller Estate Fruit Winery 116
St. Hubertus Estate Winery 74, 78
St. Laszlo Vineyards 184
St. Urban Vineyard 53
Stadler, Klaus 115
Stag's Hollow Winery 136
Stewart, Ben and Tony 89
Stewart, Richard 89–90, 165
Stuyt, John 56
Sumac Ridge Estate Winery 101, 102, 107–108, 161
Summerhill Pyramid Winery 79–80
Summerland 101
Summerland Estate Winery 102
Supernak, Frank 5, 140, 148
Sutherland, Ian and Gitta 118
Syrah 112, 128, 176

T.G. Bright & Company 168
Tait, Valerie 107, 118, 137
Tayler, Gary 53
Taylor, Ron 58, 116
Tennant, Robert 169
Tennant, Senka 169–70
Tenting 30

Thetis Island 38, 41, 42
Thetis Island Vineyards 38, 44
Thornhaven Estates 105, 106–107
Tinhorn Creek Vineyards 143, 146
Tomalty, Lorne 18
Toor Vineyards 168
Toor, Randy and Jessie 168
Township 7 Vineyards and Winery 52
Triggs, Don 176
Tugwell Creek Farm 33
Tunzelmann, Barry 66

Ulrich, Wayne and Helena 29, 30

Vancouver Island Grape Growers Association 18
Vancouver Island Vintners Association 27
Venturi, Giordano 29
Venturi-Schulze Vineyards 18, 29
Vichert, Alan 40, 41
Vicori Estate Winery 18
Victoria Estate Winery 26
Victoria Wineries 26
Vigneti Zanatta 29
Vincor International Inc. 138, 168, 176
The Vineyard at Bowen Island 38

Vintners Quality Alliance 14, 15, 61, 102
Viognier 125, 176
Violet, Claude and Inge 50–51
Visser, Kenn 179
Von Krosigk, Eric 30, 32, 102–104, 107, 121
Von Mandl, Anthony 11, 92
Von Wolff, Harry 18

Wagner, Dave 170
Wallace, Lawrence 69
Watkins, Cher and Ron 105–106
Watt, Keith 40
Watts, Tim 126
Wendenburg, Mark 112
Westham Island Estate Winery 58
Wild Goose Vineyards 138–39
Wong, Roger 74
Wyse, Jim 170–72

Yellow Point Vineyards 21

Zanatta, Dennis 27
Zanatta, Loretta 29
Zinfandel 132, 176
Zuppiger, Josef and Margrit 86
Zweigelt 86, 115

ABOUT THE AUTHOR AND PHOTOGRAPHER

JOHN SCHREINER is a writer based in North Vancouver, Canada. A native of Saskatchewan, he began touring British Columbia's wine country soon after moving to North Vancouver in 1973. An award-winning amateur winemaker, he once entertained an offer to make wine professionally until conceding that his strengths are elsewhere. Consequently, he has written six other wine books, with a primary focus on the emergence of winegrowing in British Columbia. In 2002, his work was honoured by the British Columbia industry, which gave him its annual Founders Award. He is the first wine writer to be so recognized.

KEVIN MILLER is a photographer based in Vancouver, Canada. Formerly based in Bangkok, Thailand, for six years he has travelled Southeast Asia extensively photographing the various cultures, people and architecture of the region. His specialties are architecture, travel, people, yachts, and marine photography. Upon returning to Canada, he saw the beauty of British Columbia with fresh insight and began photographing the wine regions. His work has appeared in *Popular Mechanics, Estates West, Showboats, Boat International, BC Business, Sunset*, and several in-flight magazines worldwide. His images are available at www.kevinmiller-photography.com.